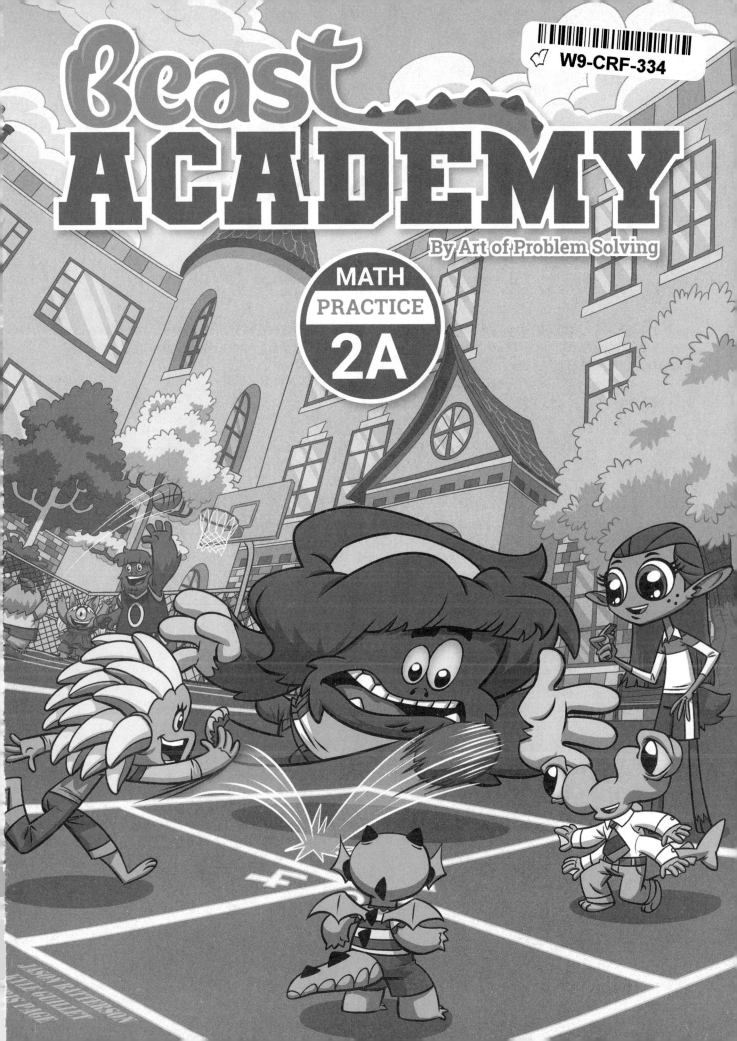

Published by: AoPS Incorporated
                   10865 Rancho Bernardo Rd Ste 100
                   San Diego, CA 92127-2102
                   info@BeastAcademy.com

ISBN: 978-1-934124-31-4

Written by Jason Batterson, Kyle Guillet, and Chris Page
Book Design by Lisa T. Phan
Illustrations by Erich Owen
Grayscales by Greta Selman

Visit the Beast Academy website at www.BeastAcademy.com.
Visit the Art of Problem Solving website at www.artofproblemsolving.com.
Printed in the United States of America.
2018 Printing

Contents:

This is Practice Book 2A in a four-book series.

2A
• Place Value
• Comparing
• Addition

2B
• Subtraction
• Expressions
• Problem Solving

2C
• Measurement
• Strategies (+&−)
• Odds & Evens

2D
• Big Numbers
• Algorithms (+&−)
• Problem Solving

For more resources and information, visit BeastAcademy.com.

This is Beast Academy Practice Book 2A.

Each chapter of this Practice book corresponds to a chapter from Beast Academy Guide 2A.

MATH PRACTICE 2A

MATH GUIDE 2A

The first page of each chapter includes a recommended sequence for the Guide and Practice books.

You may also read the entire chapter in the Guide before beginning the Practice chapter.

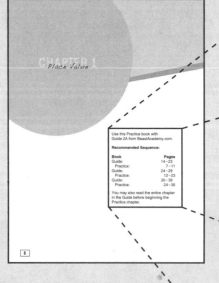

Use this Practice book with Guide 2A from BeastAcademy.com.

Recommended Sequence:

Book	Pages
Guide:	14 – 23
Practice:	7 – 11
Guide:	24 – 29
Practice:	12 – 23
Guide:	30 – 39
Practice:	24 – 35

You may also read the entire chapter in the Guide before beginning the Practice chapter.

Some problems in this book are very challenging. These problems are marked with a ★. The hardest problems have two stars!

Every problem marked with a ★ has a *hint!*

Hints for the starred problems begin on page 94.

The Sample Page

Here is where I tell you something about the problems you are doing.

EXAMPLE | Examples are included to help you understand the problems in the section.

The text that goes here usually includes an explanation of the solution to the problem given in the example.

PRACTICE | Instructions to the practice problems go here.

54. Since this problem has a star, it must be really hard! Problems with stars have hints in the back of the book, starting on page 5.

54. _____

55. After problem 54 is problem 55. All of the problems in each chapter are numbered in order, starting with problem 1. Since this problem has a pencil, your answer should include an explanation.

42 | Guide Pages: 39-43

54. **★**

42 | Guide Pages: 39-43

Some pages direct you to related pages from the Guide.

None of the problems in this book require the use of a calculator.

Solutions are in the back, starting on page 98.

A complete explanation is given for every problem!

Pretend Solutions

54. Each of the solutions in the back of this book includes a complete explanation of how to solve the problem, not just the answer. Sometimes, we will provide a solution that is different than the way you solved the problem. It is useful to read the solution to every problem even if you got the answers correct. You may learn a different way to solve the problem. Or, you may find a mistake that we made! Wouldn't it be fun to write the authors to point out a mistake they made.

55. These are just fake solutions that we wrote so that it looked like there was text here. If we put the real solutions here, it would give away some of the answers to the problems that show up in the book. Then, you wouldn't be able to have fun trying to solve them on your own! Here are some fake diagrams to go with the fake solutions:

56. If you are having trouble with a problem, it can be tempting to just look in the back of the book for the answer. The more you think about a problem, the more you learn by solving it (this is usually true even if you get the wrong answer). Do your best to figure out a problem before looking for the answer. Don't forget to use the hints for the starred problems before reading the solution.

57. Here are some more diagrams to make these fake solutions look like real ones:

58. Here is some mathy stuff:

51+52+53+50+50+50+1+2+3 = 150+6 = 156.

So, the answer is **156**. Notice that the answer is bold. This makes it easier to find the answer to each problem quickly.

59. The header below separates sections of problems. This makes it easier to find the solution you are looking for.

Fake Solutions

60. Did you notice the number on the right of the header above? That is the page number where you can find the problems in the Practice book. That way, when you are all finished reading a solution, you can flip back to the page you were working on.

61. You may be asking yourself, "Why did the author go to so much trouble writing fake text that no one can read without super-human vision?" There's no good answer, really.

62. Sometimes we give two possible ways to solve a problem. Usually, these two ways are separated / the word "or" and some lines like the ones below:

— or —

The second solution would come here. Gener ally, the quicker or more clever solution comes seond.

63. Here is another fake solution with some diagrams. It's from a different book in the Beast Acad my series. Can you figure out what is going on by looking at the diagrams?

Below is a new diagr m with the same pieces.

Can you guess what the question that goes with the diagra above was? Here's a hint: The answer is 27 squares.

64. The's another header below.

Complete Jibberish

65. The jibberish below is called Lorem Ipsum. It is complete jibberish used as placeholder text to fill the rest of the space below.

66. Curabitur velit nisl, eleifend nec interdum id, commodo id magna. Nam venenatis neque ac augue ultrices volutpat. Aenean quis diam ut ligula volutpat dapibus.

67. Makes no sense does it? I think some of those words are Latin.

68. Morbi rutrum consequat eros a rutrum. Praesent pellentesque urna ac nibh dictum consequat. Maecenas eget velit magna. Duis volutpat auctor neque sit amet congue. Aliquam erat volutpat. Suspendisse potenti. Pellentesque ut nulla lorem. Morbi consectetur purus volutpat enim vestibulum rhoncus sodales ligula aliquet.

69. Ready to get started on some math problems? Get to it!

Pretend Solutions | 111

CHAPTER 1
Place Value

Use this Practice book with Guide 2A from BeastAcademy.com.

Recommended Sequence:

You may also read the entire chapter in the Guide before beginning the Practice chapter.

 When we use more than one symbol, we always write the biggest symbols on the left: **C**'s, then **X**'s, then •'s.

When countin' coins, pirates use a • for one coin, an **X** for ten coins, and a **C** for one hundred coins.

 = •

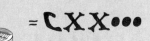 = **X**

 = **C**

 = **X**••••

 = **CXX**•••

PRACTICE | Use pirate symbols to show how many coins there are in each problem below.

1.

1. $\underline{XXXOOOOO}$

2.

2. $\underline{XXX\,OOOOOO}$

3.

3. $CXXXXOOO$

Pirate Numbers

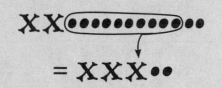

$$\text{XX} \bullet\bullet\bullet\bullet\bullet\bullet\bullet\bullet\bullet\bullet\bullet$$
$$= \text{XXX} \bullet\bullet$$

Pirates never use more than nine •'s to write a number, since ten •'s can be replaced with one X.

We also never use more than nine X's to write a number, since ten X's can be replaced with one C.

PRACTICE | Write each pirate number below using the *fewest* pirate symbols possible.

4. **X•••••••••••••** = XX○○○

5. **XXXXXXXXX•** = CX○

6. **XXXXXXXXXXX** = CXX X

7. **CXXX••••••••••••••** = CXXX X○○

8. **CCXXXXXXXXX** = CCCX○○
   •••••••••••••

9. **XXXXXXXX** = CX
★ **••••••••••••••••••**

EXAMPLE | Add the two pirate numbers below.

X X•••• + X•••••••

There are three **X**'s and twelve •'s in all.

XXX••••••••••••

We shorten this by replacing ten of the •'s with one **X**.

XXXX••

So, we have

X X•••• + X•••••••
= XXXX••

We can add pirate numbers!

PRACTICE | Add each pair of pirate numbers below.

10. X••• + ••••• = _X ooo oobooo_

11. XXXXX•• + XXX•••• = _X X X X X X X X ooooo_

12. X•••••• + XXXX* = _X X X X oo o o ooo_

Remember to write your answers with as few symbols as you can.

PRACTICE Add the pirate numbers below. Write your answer using the *fewest* pirate symbols possible.

13. **XXXX•••• + XX•••••••** = _XXXXXXXX00_

14. **XXXXX••• + XXXXX•** = _C0000_

15. **CXXXX + XXXXXXX** = _C6X_

16. **XXX••••• + XXXXXX•••••** = _CX_

17. Captain Kraken has **XXXXX** coins. He then discovers **XXXXXXX** more coins. How many coins does Captain Kraken have now?

17.  _CXX_

18. Barnacle Barney buries two treasures. One has **CX•••••** coins. The other has **XX•••••••** coins. How many total coins are buried?

18.  _CXXXX00_

PRACTICE | Write each answer below as a pirate number.

19. Captain Kraken has **XXX•••** coins. He spends **XX•** of the coins to repair the sails on his ship. How many coins does Captain Kraken have left?

19.

20. A treasure of **XXXX••••••** coins is split into two piles. If both piles contain the same number of coins, how many coins are in each pile?

20.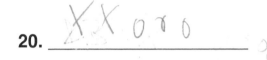

21. A treasure chest contains three bags of coins. If each bag contains **XX••••** coins, what is the total number of coins in the treasure chest?

21.

22. ★ Captain Kraken takes **XX••••** coins from a chest of **XXXXXX** coins. How many coins are left in the chest?

22. ___ X X X 0 0 0 0 0 0 ___

If you're not a pirate, you don't use •'s, X's, and ℭ's to write numbers.

0, 1, 2, 3, 4, 5, 6, 7, 8, and 9 are the ten symbols that we use to write numbers. These are called *digits*.

Digit:	Stands for:
0	zero
1	one
2	two
3	three
4	four
5	five
6	six
7	seven
8	eight
9	nine

For example, 43 is a two-digit number that uses the digits 4 and 3.

PRACTICE | Write each pirate number below as a *two-digit number*.

23. X••• = 13

24. X•••••••• = 18

25. XX•••• = 24

26. XXXXX• = 61

27. XXXXX••• = 54

28. XXXX••••• = 45

29. XX••• = 23

30. XXXXX = 50

two hundreds · five tens · eight ones

ᒐᒐ XXXXX ·········

five tens
two hundreds ⟶ ⟵ eight ones
= 258

In a pirate number, the ᒐ's, X's, and •'s let us know how many hundreds, tens, and ones there are.

In a three-digit number, the digits tell us how many hundreds, tens, and ones there are.

PRACTICE | Write each pirate number as a ***three-digit number***.

31. ᒐXXXX••• = 143

32. ᒐX········ = 118

33. ᒐᒐᒐXX• = 321

34. ᒐXX•••• = 124

35. ᒐᒐᒐᒐᒐX•• = 512

36. ᒐ••••• = 105

37. ᒐXXX = 130

38. ᒐᒐᒐᒐᒐᒐᒐ = 700

A digit's location in a number is called its **place value**. Every three-digit number has a hundreds place, a tens place, and a ones place.

In 309, we say that 3 is the hundreds digit, 0 is the tens digit, and 9 is the ones digit.

hundreds tens ones

309

PRACTICE | Answer each question below about place value.

39. Circle every two-digit number below.

255   (55)   (25)   522   555   (52)

40. Circle every number below that has a 2 in the hundreds place.

(209)   902   92   29   (290)   920

41. Circle every number below that has tens digit 7.

(477)   747   (774)   447   (474)   744

42. Circle every number below whose hundreds digit is larger than its ones digit.

(978)   798   879   (897)   789   (987)

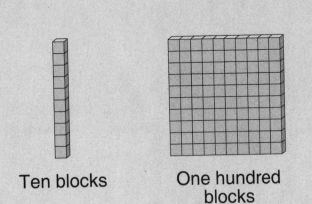
Counting Blocks
PLACE VALUE

One block   Ten blocks   One hundred blocks

EXAMPLE | Write the number of blocks shown below.

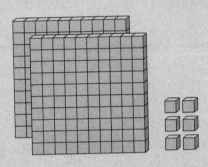

There are 2 hundreds, 0 tens, and 6 ones.
So, there are **206** blocks.

PRACTICE | Write the number of blocks in each problem below.

57.

58.

57. _____

58. _____

We never write 0 as the leftmost digit of a number with more than one digit.

For example, we always write 7 tens and 3 ones as 73, never as "073."

PRACTICE | Answer each question below about place value.

43. Arrange the digits in 322 to write a new three-digit number that has a 3 in the tens place.

43. _232_

44. Arrange the digits in 750 to write a new three-digit number that has ones digit 7.

44. _507_

45. How many different two-digit numbers have 0 as a digit?

45. _9_

46. Write three **different** three-digit numbers that use the digits 7, 8, and 8.

46. _788_, _878_, _887_

47. Use the digits 5, 7, and 9 to write a number whose ones digit is larger than its tens digit, but smaller than its hundreds digit.

47. _975_

In a **Number Search**, we circle 3-digit numbers in a row of digits. The 3-digit numbers cannot overlap each other.

EXAMPLE | In the row of digits below, circle two *different* 3-digit numbers that have 6 in the tens place.

2 6 6 2 6 6 2

Below is the only way to circle two different 3-digit numbers that have tens digit 6 and do not overlap.

2 6 6 2 6 6 2

PRACTICE | Solve each Number Search below by circling numbers that do not overlap.

48. Circle a 3-digit number with ones digit 2 and hundreds digit 4.

2 3 4 2 4 3 2

49. Circle two *different* 3-digit numbers that have a 5 in the tens place.

5 5 4 5 5 4 5 5

50. Circle two *different* 3-digit numbers that have a 9 in the ones place.

9 9 9 8 9 9 9

51. Circle two *different* 3-digit numbers that have a 0 in the ones place.

0 0 1 1 0 0 1 1 0

Remember, the numbers you circle in these problems can't overlap.

PRACTICE | Solve each Number Search below by circling numbers that do not overlap.

52. Circle three *different* 3-digit numbers that have a 0 in the ones place.

0 1 2 0 0 2 2 0 0 1 2 0

53. Circle three copies of the *same* 3-digit number.

1 2 3 1 4 2 1 3 1 4 1 2 3 1 4 2

54. Circle three 3-digit numbers that have their largest digit in the tens place.

8 7 6 5 7 6 5 4 6 5 4 3 5 4 3 2

55. ★ Circle four 3-digit numbers that all have the same tens digit.

4 3 2 1 1 2 3 4 3 1 2 4 4 2 1 3

56. ★ Circle four *different* 3-digit numbers.

7 3 3 7 3 3 7 7 3 3 7 3 3 7 3

We never write 0 as the leftmost digit of a number with more than one digit.

For example, we always write 7 tens and 3 ones as 73, never as "073."

PRACTICE | Answer each question below about place value.

43. Arrange the digits in 322 to write a new three-digit number that has a 3 in the tens place.

43. 232

44. Arrange the digits in 750 to write a new three-digit number that has ones digit 7.

44. 507

45. How many different two-digit numbers have 0 as a digit?

45. 9

46. ★ Write three *different* three-digit numbers that use the digits 7, 8, and 8.

46. 788 , 878 , 887

47. ★ Use the digits 5, 7, and 9 to write a number whose ones digit is larger than its tens digit, but smaller than its hundreds digit.

47. 975

In a **Number Search**, we circle 3-digit numbers in a row of digits. The 3-digit numbers cannot overlap each other.

EXAMPLE | In the row of digits below, circle two *different* 3-digit numbers that have 6 in the tens place.

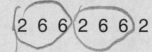

Below is the only way to circle two different 3-digit numbers that have tens digit 6 and do not overlap.

PRACTICE | Solve each Number Search below by circling numbers that do not overlap.

48. Circle a 3-digit number with ones digit 2 and hundreds digit 4.

2 3 4 2 4 3 2

49. Circle two *different* 3-digit numbers that have a 5 in the tens place.

5 5 4 5 5 4 5 5

50. Circle two *different* 3-digit numbers that have a 9 in the ones place.

9 9 9 8 9 9 9

51. Circle two *different* 3-digit numbers that have a 0 in the ones place.

0 0 1 1 0 0 1 1 0

Remember, the numbers you circle in these problems can't overlap.

PRACTICE | Solve each Number Search below by circling numbers that do not overlap.

52. Circle three **different** 3-digit numbers that have a 0 in the ones place.

0 1 2 0 0 2 2 0 0 1 2 0

53. Circle three copies of the **same** 3-digit number.

1 2 3 1 4 2 1 3 1 4 1 2 3 1 4 2

54. Circle three 3-digit numbers that have their largest digit in the tens place.

8 7 6 5 7 6 5 4 6 5 4 3 5 4 3 2

55. Circle four 3-digit numbers that all have the same tens digit.
★

4 3 2 1 1 2 3 4 3 1 2 4 4 2 1 3

56. Circle four **different** 3-digit numbers.
★

7 3 3 7 3 3 7 7 3 3 7 3 3 3 7 3

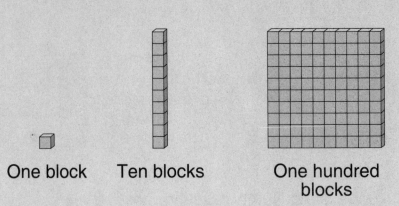

One block   Ten blocks   One hundred blocks

EXAMPLE | Write the number of blocks shown below.

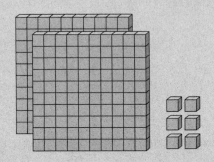

There are 2 hundreds, 0 tens, and 6 ones.
So, there are **206** blocks.

PRACTICE | Write the number of blocks in each problem below.

57.

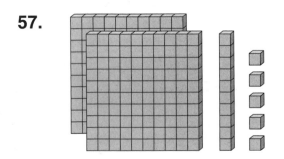

58.

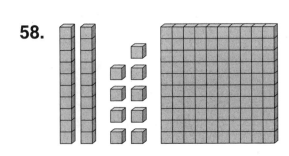

57. _____

58. _____

We can combine 10 ones to make 1 ten, or 10 tens to make 1 hundred.

For these problems, you need to arrange some blocks into larger groups.

PRACTICE | Write the number of blocks in each problem below.

59.

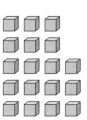

59. _____

60.

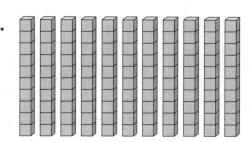

60. _____

61.

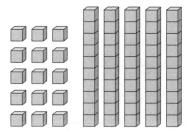

61. _____

62.

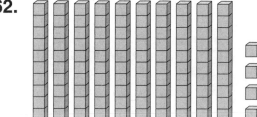

62. _____

63.

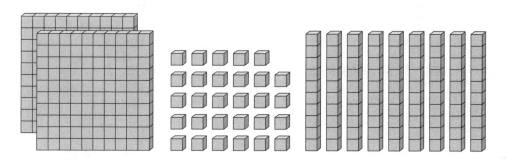

63. _____

EXAMPLE | What two-digit number is the same as 3 tens and 13 ones?

Every place value must contain a single digit. We cannot write a number with 13 in the ones place. So, we need to **regroup**. We can regroup 10 ones to make 1 ten.

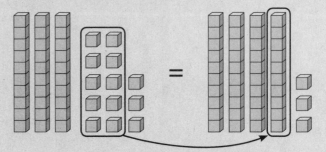

So, 3 tens and 13 ones is the same as 4 tens and 3 ones, which is **43**.

PRACTICE | Answer each question below.

64. What two-digit number is the same as 7 tens and 17 ones?   64. _____

65. What two-digit number is the same as 3 tens and 28 ones?   65. _____

66. What three-digit number is the same as 5 hundreds, 2 tens, and 11 ones?   66. _____

67. What three-digit number is the same as 15 tens?

PRACTICE | Fill the blanks to complete each statement below.

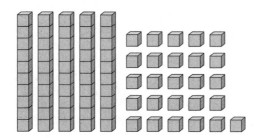

68. 5 tens and 26 ones is the same as __6__ tens and _____ ones.

69. 5 tens and 26 ones is the same as __7__ tens and _____ ones.

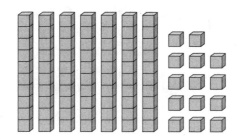

70. 7 tens and 14 ones is the same as __8__ tens and _____ ones.

71. 7 tens and 14 ones is the same as __6__ tens and _____ ones.

Try these last three without using blocks.

72. 12 tens is the same as _____ hundred and __2__ tens.

73. ★ 3 hundreds and 7 tens is the same as _____ tens.

74. ★★ 222 is the same as _____ tens and __12__ ones.

Regrouping

This chart shows five different ways to group 74 into tens and ones.

What is the missing number in this chart?

tens	ones
7	4
6	14
0	74
2	54
5	

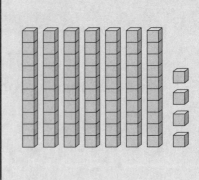

74 means 7 tens and 4 ones.

We can **break** a ten to make 10 ones.
To leave 5 tens, we break 2 of the 7 tens.
Breaking 2 tens makes 20 ones.

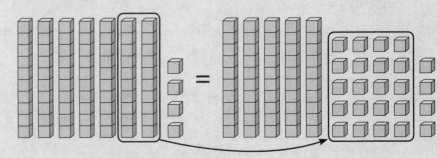

So, 74 is 5 tens and 20+4 = **24** ones.

PRACTICE | Fill the missing entries in each chart below so that each row describes the number of blocks to its right.

75.

tens	ones
6	
5	
4	
3	

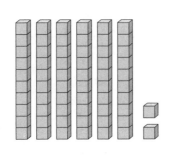

76.

tens	ones
	7
3	
	27
1	

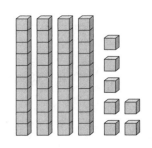

PRACTICE | Fill the missing entries in each chart below so that each row describes the amount above the chart.

39

77.

tens	ones
0	
	29
2	
	9

42

78.

tens	ones
4	
3	
	22
1	

58

79.

tens	ones
	48
	18
3	
5	

50

80.

tens	ones
	0
4	
3	
	40

91

81.

tens	ones
7	
	81
	71
	1

77

82.

tens	ones
	17
1	
	7
0	

83. Fill each blank.

890 = _____ tens

605 = _____ tens and 5 ones

534 = _____ tens and 34 ones

84. Fill each blank with a 3-digit number.

17 tens = _____

90 tens = _____

4 hundreds and 44 tens = _____

We can use what we know about place value to help us add one, ten, or one hundred to a number.

EXAMPLE Add: 276+1
276+10
276+100

276+1: Adding 1 to 276 increases its ones digit by 1. So, 276+1 = **277**.

276+10: Adding 10 to 276 increases its tens digit by 1. So, 276+10 = **286**.

276+100: Adding 100 to 276 increases its hundreds digit by 1. So, 276+100 = **376**.

PRACTICE Solve each addition problem below.

85. 134+1 = _____

86. 134+10 = _____

87. 134+100 = _____

88. 555+1 = _____

89. 555+10 = _____

90. 555+100 = _____

91. 768+1 = _____

92. 768+10 = _____

93. 768+100 = _____

94. 987+11 = _____

95. 798+101 = _____

96. 879+110 = _____

97. 729+11 = _____

98. 279+101 = _____

99. ★ 792+110 = _____

EXAMPLE | Subtract: 427−1
427−10
427−100

We can also subtract 1, 10, or 100 from a number using what we know about place value.

427−1: Subtracting 1 from 427 decreases its ones digit by 1. So, 427−1 = **426**.

427−10: Subtracting 10 from 427 decreases its tens digit by 1. So, 427−10 = **417**.

427−100: Subtracting 100 from 427 decreases its hundreds digit by 1. So, 427−100 = **327**.

PRACTICE | Solve each subtraction problem below.

100. 615−1 = _____

101. 615−10 = _____

102. 615−100 = _____

103. 888−1 = _____

104. 888−10 = _____

105. 888−100 = _____

106. 432−1 = _____

107. 432−10 = _____

108. 432−100 = _____

109. 543−11 = _____

110. 354−101 = _____

111. 435−110 = _____

112. 120−1 = _____

113. 102−100 = _____

114. 201−10 = _____

EXAMPLE | Add 498+10+10.

Since 490 is 49 tens, 498 is 49 tens and 8 ones.

Adding 1 ten to 498 gives 50 tens and 8 ones, which is 508.

Adding 1 more ten gives 51 tens and 8 ones, which is 518.

So, 498+10+10 = **518**.

PRACTICE | Fill the blanks to complete each pattern below.

115.
$$\overset{+10}{\frown} \overset{+10}{\frown} \overset{+10}{\frown} \overset{+10}{\frown} \overset{+10}{\frown} \overset{+10}{\frown} \overset{+10}{\frown} \overset{+10}{\frown}$$
130 , 140 , 150 , _____ , _____ , _____ , _____ , _____ , _____ .

116.
$$\overset{+10}{\frown} \overset{+10}{\frown} \overset{+10}{\frown} \overset{+10}{\frown} \overset{+10}{\frown} \overset{+10}{\frown} \overset{+10}{\frown} \overset{+10}{\frown}$$
255 , 265 , 275 , _____ , _____ , _____ , _____ , _____ , _____ .

117.
$$\overset{+10}{\frown} \overset{+10}{\frown} \overset{+10}{\frown} \overset{+10}{\frown} \overset{+10}{\frown} \overset{+10}{\frown} \overset{+10}{\frown} \overset{+10}{\frown}$$
_____ , _____ , 562 , 572 , 582 , _____ , _____ , _____ , _____ .

118.
★
$$\overset{+10}{\frown} \overset{+10}{\frown} \overset{+10}{\frown} \overset{+10}{\frown} \overset{+10}{\frown} \overset{+10}{\frown} \overset{+10}{\frown} \overset{+10}{\frown}$$
_____ , _____ , _____ , _____ , _____ , _____ , 424 , _____ , _____ .

EXAMPLE | Subtract 207 − 10.

2 hundreds, 0 tens, and 7 ones is the same as
1 hundred, 10 tens, and 7 ones.

Taking away 1 ten leaves us with
1 hundred, 9 tens, and 7 ones, which is **197**.

– or –

2 hundreds is the same as 20 tens. So, 207 is
20 tens and 7 ones. Taking away 1 ten leaves
19 tens and 7 ones, which is **197**.

PRACTICE | Answer each question below.

119. Fill each blank with the correct answer:

800 − 10 = _____   703 − 10 = _____   309 − 10 = _____

120. There are 506 chipskunks in a field. After 10 chipskunks   120. _____
run away, how many chipskunks are left in the field?

121. Captain Kraken's secret treasure is 405 lollipops.   121. _____
His crew eats 10 of them. How many lollipops are
left in his treasure?

122. Grogg's number is 1 less than Lizzie's.   122. _____
★ Lizzie's number is 10 less than Alex's.
Alex's number is 100 less than Winnie's.
Winnie's number is 209. What is Grogg's number?

In a **Digit Difference Grid**, numbers in squares that share a side must be 1, 10, or 100 apart. For example, a square next to 355 can be filled with any of the six bold numbers below.

$$355 + 1 = \mathbf{356} \qquad 355 - 1 = \mathbf{354}$$

$$355 + 10 = \mathbf{365} \qquad 355 - 10 = \mathbf{345}$$

$$355 + 100 = \mathbf{455} \qquad 355 - 100 = \mathbf{255}$$

A number can only be used once in a Digit Difference Grid.

EXAMPLE | Fill the empty square to complete the following Digit Difference Grid.

354	355
	255

To get from 354 to 255, we subtract 100 and add 1, or add 1 and subtract 100.

So, the number in the empty square is either 354 − 100 = 254, or 354 + 1 = 355.

Each number can only be used once. 355 is already in the grid, so the empty square must be **254**.

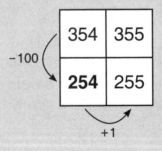

PRACTICE | Fill the empty squares to complete each Digit Difference Grid below.

123.

63		65

124.

366		566

125.

155		175

PRACTICE | Fill the empty squares to complete each Digit Difference Grid below.

126. | 87 | | 85 |

127. | 443 | | 243 |

128. | 289 | | 269 |

129. | 593 | | 613 |

130. | 101 | | 99 |

131. | 402 | | 382 |

132. | 222 | | 22 |

133. ★ | 62 | | 82 | 182 | | 83 |

Remember, you can only use a number once in each Digit Difference Grid.

134. ★ | 91 | | 100 | 90 | | 99 |

PRACTICE | Fill the empty squares to complete each Digit Difference Grid below.

135.

17	
27	28

136.

833	832
	732

137.

5	105
15	

138.

835	825
	824

139.

179	
189	89

140.

	199
210	209

141.

9	10	
	20	120

142.

	66	67
75	76	

143.

285	286	296
	386	

PRACTICE | Fill the empty squares to complete each Digit Difference Grid below.

144.

46	56	66
45		

145.

172	272	262
173		

146.

329	339	340
		240

147.

		706
586		606

148.

777		797
		798

149.

	89	
100		80

150. ★

	13	23
103		

151. ★

114		24
	125	

152. ★★

		66
	146	

We can add or subtract more than 1 one, 1 ten, or 1 hundred at a time.

EXAMPLE | Add 427+60.

60 is 6 tens. So, adding 60 is the same as adding 6 tens.

427 is 4 hundreds, 2 tens, and 7 ones.

Adding 6 tens to 427 gives
4 hundreds, 8 tens, and 7 ones, which is **487**.

Adding 6 tens to a number increases its tens digit by 6, unless we have to regroup.

PRACTICE | Answer each addition and subtraction problem below.

153. 213+500 = _____

154. 213+5 = _____

155. 213+50 = _____

156. 586−4 = _____

157. 586−400 = _____

158. 586−40 = _____

159. 121+70 = _____

160. 121+7 = _____

161. 121+700 = _____

162. 475−300 = _____

163. 475−30 = _____

164. 475−3 = _____

For these problems you'll need to break or regroup.

PRACTICE | Answer each addition and subtraction problem below. Write your answers using digits.

165. Add:

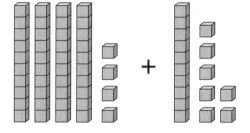

165. $44 + 17 =$ _____

166. Subtract:

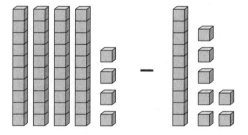

166. $44 - 17 =$ _____

167. Add ✗✗✗✗•• + ✗✗✗✗✗✗••••

167. $42 + 64 =$ _____

168. Subtract ⸂✗•• − ✗✗✗✗

168. $112 - 40 =$ _____

169. Add 7 tens to 683.

169. $683 + 70 =$ _____

170. Subtract 8 tens from 446.

170. $446 - 80 =$ _____

PRACTICE | Answer each problem below.

171. What number is 111 more than 789?

171. _____

172. Write the four different three-digit numbers that use the digits 0, 1, and 2 once each.

_____, _____, _____, _____

173. How many symbols are in the pirate number for 789?
Review pirate numbers on page 7.

173. _____

174. ★ There are only three 3-digit numbers that can be written using only two pirate symbols. Write them as pirate numbers.
Review pirate numbers on page 7.

_____, _____, _____

PRACTICE | Answer each problem below.

175. Fill the blanks to complete the pattern below.
★

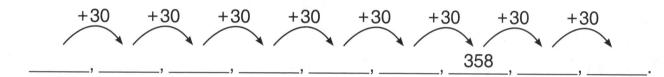

_____, _____, _____, _____, _____, _____, 358 , _____, _____.

176. What three-digit number is 77 tens plus 77 ones?

176.

177. Adding a number's digits gives you the number's ***digit sum***. For example, 345 has a digit sum of $3+4+5=12$. What is the smallest three-digit number that has a digit sum of 12?

177.

178. If you write every number from 1 to 100, how many times will you write the digit 5?
★
★

178. _____

CHAPTER 2
Comparing

Use this Practice book with
Guide 2A from BeastAcademy.com.

Recommended Sequence:

Book	Pages:
Guide:	42-55
Practice:	37-49
Guide:	56-67
Practice:	50-63

You may also read the entire
chapter in the Guide before
beginning the Practice chapter.

A *number line* shows numbers in order from left to right.

Tick marks on the number line stand for whole numbers.

PRACTICE | Label the missing numbers on each number line below.

1.

0 1 ☐ 3 4 5 6 7 ☐ 9 10

2.

0 ☐ 2 3 4 5 6 ☐ 8 9 10 11 ☐ 13 14

3.

☐ ☐ 15 18 ☐

4.

☐ 30 ☐ 35 ☐

EXAMPLE | Label 50, 60, and 63 on the number line below.

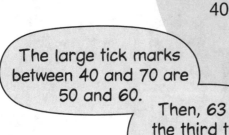

The large tick marks between 40 and 70 are 50 and 60.

Then, 63 is the third tick mark to the right of 60.

PRACTICE | Draw arrows as shown in the example above to label the numbers given.

5. Label 16, 20, 25, and 43.

6. Label 61, 75, 84, and 90.

7. Label 34, 45, 56, and 67.

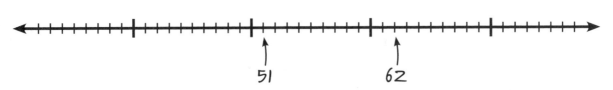

Number lines don't always have a tick mark for **every** whole number.

PRACTICE | Answer each number line question below.

8. Label the missing numbers in the boxes below.

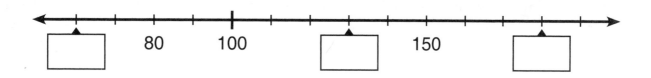

80   100   150

9. Label the missing numbers in the boxes below.

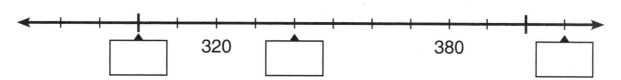

320   380

10. Label the missing numbers in the boxes below.

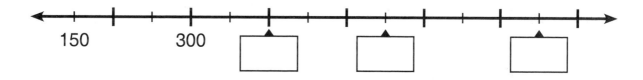

150   300

11. ★ Label each arrow with a three-digit number that includes the digits 2, 4, and 6 once each.

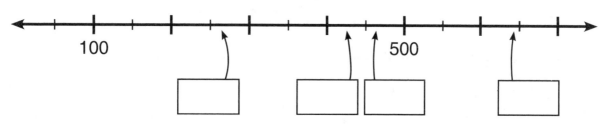

100   500

COMPARING

EXAMPLE | How far is 37 from 62 on the number line?

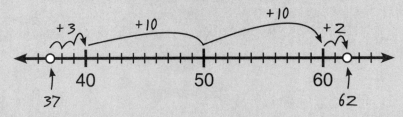

From 37 to 40 is 3 units.
From 40 to 60 is 10+10 = 20 units.
From 60 to 62 is 2 units.

So, the distance from 37 to 62 is
3+20+2 = **25** units.

> We can find the distance between two numbers by counting up on the number line.

PRACTICE | Find the distance between each pair of numbers below.

12. How far is 19 from 60 on the number line?   12. _____

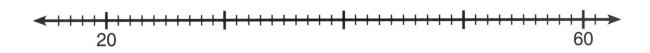

13. What is the distance between the two dots on the number line below?   13. _____

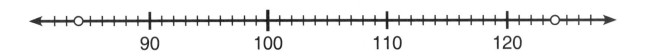

14. Find the distance from 54 to 81 on the number line.   14. _____

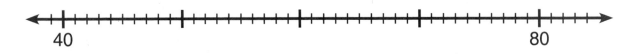

EXAMPLE | Fill in the blank: 58 and ____ are the same distance from 70 on the number line.

The distance from 58 to 70 is $2+10=12$ units.

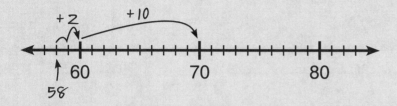

We say that 70 is **halfway** between 58 and 82.

To find the other number that is 12 units from 70, we count up 12 units from 70.

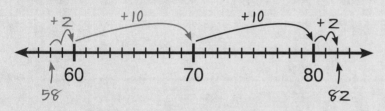

So, 58 and **82** are the same distance from 70.

PRACTICE | Fill each blank below.

15. 18 and ____ are the same distance from 25 on the number line.

16. 33 and ____ are the same distance from 50 on the number line.

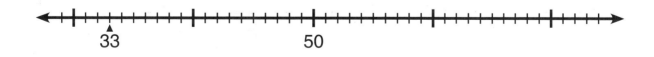

17. 98 and ____ are the same distance from 73 on the number line.

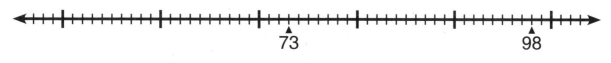

EXAMPLE | What number is halfway between 15 and 55?

We can count in from 15 and 55. As long as we count in by the same amount, the number halfway between our points stays the same.

The number halfway between 15 and 55 is the same as the number halfway between 20 and 50.

We keep counting in by equal amounts until we reach the middle.

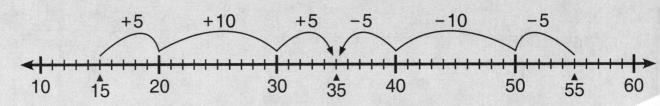

So, **35** is halfway between 15 and 55.

PRACTICE | Answer each question below.

18. What number is halfway between 45 and 65?

18. _____

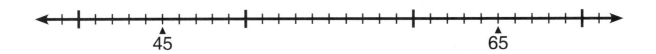

19. What number is halfway between 210 and 260?

19. _____

20. What number is halfway between 27 and 73?

20. _____

PRACTICE | Answer each question below.

21. What number is halfway between 32 and 76?

21. _____

⭐

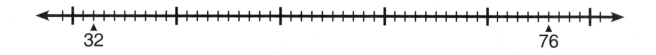

32   76

22. What number is halfway between 534 and 584?

22. _____

⭐

534   584

Try solving the problems below **without** a number line!

23. What number is halfway between 70 and 130?

23. _____

24. What number is halfway between 118 and 162?

24. _____

25. What number is halfway between 43 and 89?

25. _____

⭐

26. Is 64 closer to 40 or to 90?

26. _____

⭐

27. Is 385 closer to 350 or to 410?

27. _____

⭐

Number Line Conquest is a game for two players.

Start with an unlabeled number line. Players take turns labeling tick marks as "bases" until each player has three bases.

After all six bases are labeled, the game is scored. Players get 1 point for each blank tick mark that is closer to one of their bases than to their opponent's.

The player with the most points wins.

EXAMPLE | Who wins the game of Number Line Conquest between Winnie (W) and Grogg (G) below?

Winnie labels each blank tick mark that is closer to one of her bases with a w. Grogg labels each blank tick mark that is closer to one of his bases with a g. The tick marks that are the same distance from both players' bases are not scored for either player.

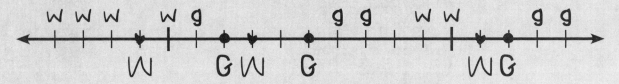

Winnie scores 6 points and Grogg scores 5. So, **Winnie wins**.

PRACTICE | Find a partner and play Number Line Conquest on the number lines below. Take turns going first.

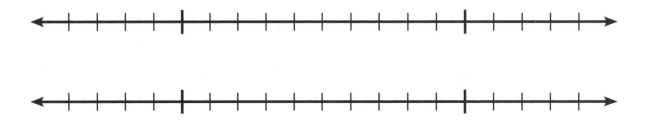

Print more game boards at BeastAcademy.com.

PRACTICE | Place a third X on each game board below so that Player X wins the game of Number Line Conquest.

28.

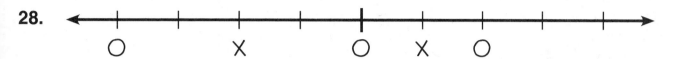

29.
★

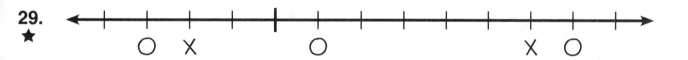

30.
★

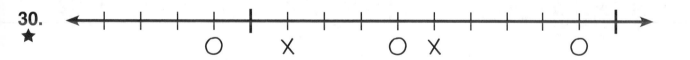

31.
★

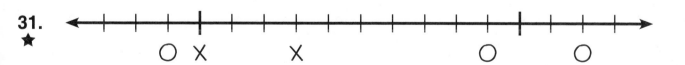

32.
★

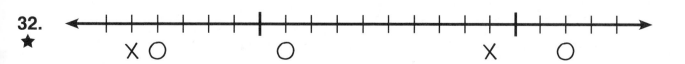

In a **Honeycomb Path** puzzle, the goal is to fill every empty hexagon (⬡) with a number so that a path of **consecutive numbers** crosses every hexagon.

EXAMPLE | Fill the empty hexagons to solve the Honeycomb Path puzzle below.

Consecutive numbers come one after another. For example, 11, 12, and 13 are consecutive.

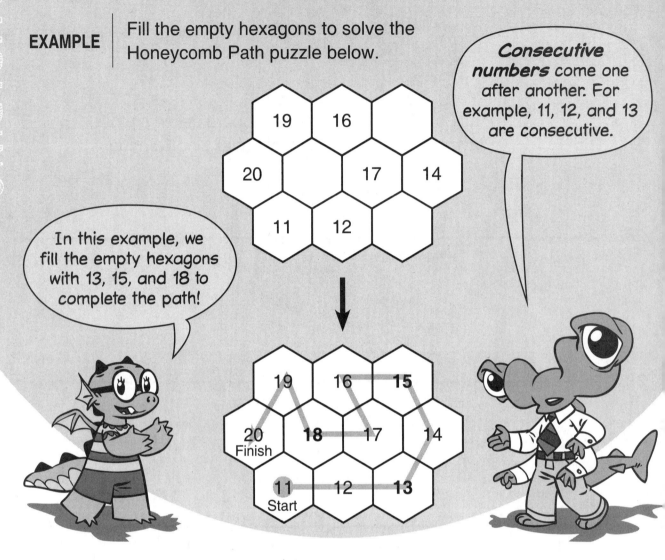

In this example, we fill the empty hexagons with 13, 15, and 18 to complete the path!

PRACTICE | Fill the empty hexagons with numbers to solve each Honeycomb Path puzzle below.

33.

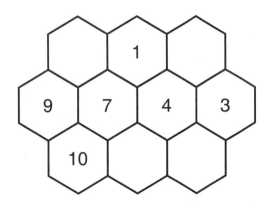

34.

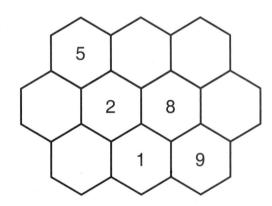

PRACTICE | Fill the empty hexagons with numbers to solve each Honeycomb Path puzzle below.

35.

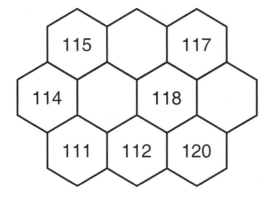

36.

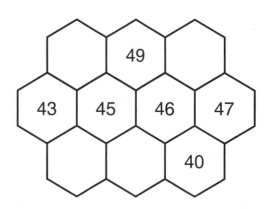

37.

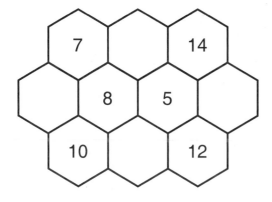

38.

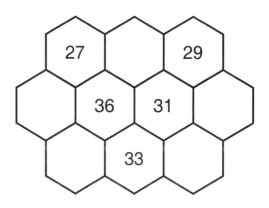

39.

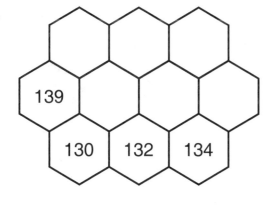

40.

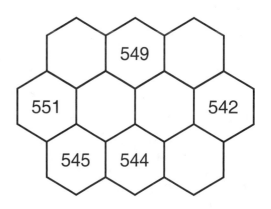

PRACTICE | Fill the empty hexagons with numbers to solve each Honeycomb Path puzzle below.

41.

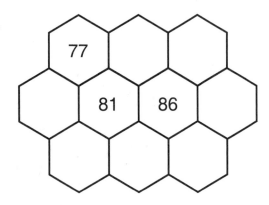

42.

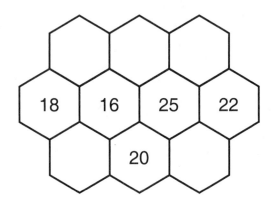

43.
★

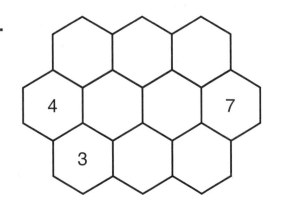

44.
★

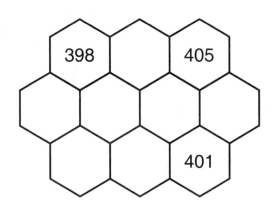

45.
★

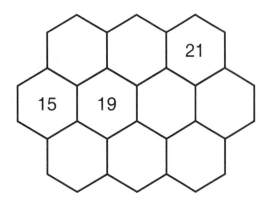

46.
★

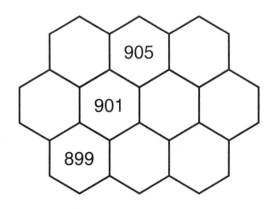

PRACTICE | Fill the empty hexagons with numbers to solve each Honeycomb Path puzzle below.

47.

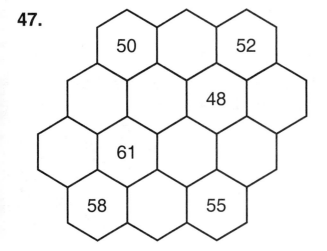

48.
★

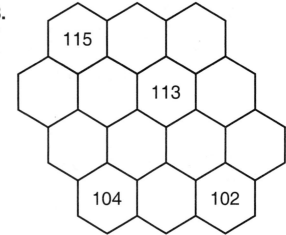

49.
★

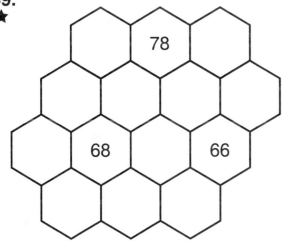

50.
★

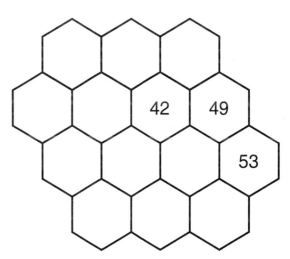

Print more of these puzzles at BeastAcademy.com.

An equals sign (=) shows that two amounts are the same value. For example,

$$3+3=6.$$

If two amounts are not equal, we use < or > to show which is larger.

The < symbol means "is less than."
The > symbol means "is greater than."

For example,

6 < 7 is read "six is less than seven," and
100 > 99 is read "100 is greater than 99."

We have symbols to show when one value is larger than another.

The < and > symbols always "eat" the bigger amount!

$100 > 99$

PRACTICE | Fill each circle below with <, >, or = .

51. 98 ◯ 89

52. 2+7 ◯ 27

53. 9+3 ◯ 3+9

54. 40+4 ◯ 50−5

55. 213 ◯ 132

56. 7+80 ◯ 70+8

57. 19 ones ◯ 2 tens

58. 30 tens ◯ 3 hundreds

59. 199+202 ◯ 200+200

60. 99+99+99 ◯ 300−3

Fill each blank below with a digit.

For example, 5|8 is the number 58.

PRACTICE | Answer each question below.

61. Fill the blank with a digit to make the comparison below true.

$$6\boxed{}<61$$

62. Fill the blank with a digit to make the comparison below true.

$$\boxed{}4>85$$

63. Use the digits *2*, *3*, and *4* once each to make all three comparisons true.

$$30<\boxed{}1$$

$$7\boxed{}>73$$

$$\boxed{}3<34$$

64. Use the digits *5*, *6*, *7*, and *8* once each to make all three comparisons true.
★

$$\boxed{}1>62$$

$$7\boxed{}<\boxed{}5$$

$$46>4\boxed{}$$

EXAMPLE | Order the numbers 87, 877, 787, and 778 from least to greatest.

87 is the only number that is less than 100, so 87 is the smallest.

87, ___, ___, ___

Next, we have two numbers in the 700's and one number in the 800's. Any number in the 800's is larger than any number in the 700's, so 877 is largest.

87, ___, ___, 877

Finally, since 78 is less than 87, we know 778 is less than 787. So, from least to greatest, we have

87, 778, 787, 877.

PRACTICE | Solve each problem below.

65. Order the numbers 32, 233, 323, and 223 from *least to greatest*.

_____, _____, _____, _____

66. Order the numbers 714, 471, 741, and 417 from *least to greatest*.

_____, _____, _____, _____

67. What is the greatest three-digit number whose digits are all different?

67. _____

68. What is the smallest three-digit number whose digits are all different?

68. _____

EXAMPLE | Order the numbers below from greatest to least.

When comparing more than a few numbers, it helps to organize them by place value.

Hundreds matter more than tens, and tens matter more than ones.

87, 11, 203, 96, 451, 8, 112

Hundreds	Tens	Ones
	8	7
	1	1
2	0	3
	9	6
4	5	1
		8
1	1	2

203, 451, and 112 are the only numbers with a hundreds digit. 451 has the largest hundreds digit, followed by 203, then by 112. So, the first three numbers in our list are 451, 203, and 112.

87, 11, and 96 are the only remaining numbers with a tens digit. From greatest to least, we have 96, 87, and 11.

The smallest number is 8.

So, from greatest to least, we have

451, 203, 112, 96, 87, 11, 8.

PRACTICE | In the problems below, fill the blanks to order the numbers from **greatest to least**.

69. 785 _____
115 _____
51 _____
23 _____
6 _____
758 _____
511 _____
203 _____

70. 45 _____
345 _____
435 _____
54 _____
4 _____
543 _____
5 _____
53 _____

71. 11 _____ Greatest
111 _____
919 _____
91 _____
99 _____
119 _____
9 _____
19 _____ Least

In a **Number Path** puzzle, the goal is to trace the path that crosses all of the numbers in the grid from least to greatest.

EXAMPLE | Trace the path that connects the numbers in the grid below in order from least to greatest.

121	98	92	221
124	97	95	220
125	152	159	219
127	151	210	212

→

121	98	92	221
124	97	95	220
125	152	159	219
127	151	210	212

We start at the smallest number, 92...

...then connect numbers from least to greatest until the path has crossed every number.

PRACTICE | In each puzzle below, trace the path that connects the numbers in the grid in order from least to greatest.

72.

24	21	17	16
25	74	77	14
29	71	70	68
35	36	63	64

73.

97	95	89	82
98	99	78	80
35	36	77	75
33	37	73	74

PRACTICE | In each puzzle below, trace the path that connects the numbers in the grid in order from least to greatest.

74.

995	509	550	559
990	505	500	590
959	909	905	595
955	950	900	599

75.

465	534	536	543
456	453	564	546
435	436	345	346
365	364	356	354

76.

41	43	411	414	417
50	47	79	77	441
57	70	74	75	447
717	714	711	707	471
741	745	749	477	474

77.

886	885	65	62	59
868	881	68	83	56
862	655	650	88	53
858	668	561	516	518
851	686	559	553	551

78.

511	195	159	155	885
515	188	181	151	881
518	551	115	118	858
558	555	811	815	855
581	585	588	819	851

79.

468	482	486	624	628
462	428	426	648	642
284	286	288	682	684
268	246	862	864	824
264	248	846	842	826

PRACTICE | Solve each problem below.

80. Use the blanks to order the following numbers from *least to greatest*.

72   27   16   108   61   18

_____ < _____ < _____ < _____ < _____ < _____

81. Use the blanks to order the following numbers from *greatest to least*.

65   506   650   560   605   56

_____ > _____ > _____ > _____ > _____ > _____

82. **How many** different whole numbers could replace the gray box below to make a true statement?

82. _____

$$10 < \blacksquare < 20$$

PRACTICE | Solve each problem below.

83. The eight numbers below use only the digits 2 and 9. Use the blanks to order these numbers from least to greatest.

29   99   222   9   22   2   92   229

_____ < _____ < _____ < _____ < _____ < _____ < _____ < _____

84. Fill each blank below with a number using only the digits 3 and 4.

33 < _____ < 43 < 44 < _____ < 334 < _____ < _____ < 433

85. ★ There are six different numbers that can be made using the digits 1, 2, and 3 exactly once each. Two of the numbers are shown below. Fill the remaining four blanks so that all numbers are in order from least to greatest.

123 < _____ < _____ < _____ < 312 < _____

86. ★ The statement below lists the smallest eight numbers that can be written using only the digits 0 and 5. Fill in the four missing numbers.

0 < 5 < 50 < _____ < _____ < _____ < _____ < 555

87. ★ The statement below lists the smallest eight numbers that can be written using only the digits 7 and 8. Fill in the four missing numbers.

7 < 8 < 77 < _____ < _____ < _____ < _____ < 778

EXAMPLE | Fill the blanks below with the digits 4, 6, and 8.

4☐ < ☐7 < 4☐

The first and last number each have tens digit 4...

...so the middle number also has tens digit 4.

4☐ < **4**7 < 4☐

4**6** < **4**7 < 4**8**

We fill the remaining blanks with 6 and 8.

PRACTICE | In each problem below, use the given digits once each to fill the blanks.

88. **Digits:** 4, 7, 8

☐2 < ☐2 < ☐2

89. **Digits:** 3, 5, 9

☐ < ☐0 < ☐0 < 80

90. **Digits:** 2, 4, 6

34 < 3☐ < 4☐ < 4☐

91. **Digits:** 4, 5, 6, 7

☐☐ < 4☐ < ☐4 < 60

PRACTICE | In each problem below, use the given digits once each to fill the blanks.

92. **Digits**: 6, 8, 9

$$\square 9 < 9 \square < \square 7$$

It might help to cross out the digits as you use them!

93. **Digits**: 5, 6, 7

$$5 < \square < \square 5 < \square 5 < 75$$

94. **Digits**: 1, 1, 1, 1, 2, 2, 2, 2

$$\square\square < \square\square < \square\square < \square\square$$

95. ★ **Digits**: 3, 4, 5, 5, 9, 9

$$40 < \square\square < \square\square < \square\square < 70$$

96. ★ **Digits**: 5, 5, 5, 5, 7, 7, 9, 9, 9

$$\square\square < \square 7 < \square\square < 7\square < \square 7 < \square\square$$

PRACTICE | Answer each question below.

97. ★ What number is halfway between 123 and 321?

97. _____

98. ★ Label the missing numbers in the boxes below.

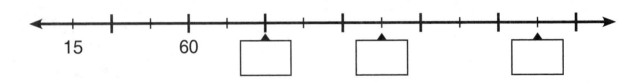

15   60

99. ★ Alex, Lizzie, and Grogg each pick a number on the number line. Lizzie's number is halfway between Alex's number and Grogg's number. If Alex picked 25 and Lizzie picked 55, what number did Grogg pick?

99. _____

PRACTICE | Answer each question below.

100. ★ Ms. Q marks the number 50 on the number line.
Alex marks the number 15 units to the left of Ms. Q's number.
Grogg marks the number 25 units to the right of Alex's number.
Lizzie marks the number 35 units to the left of Grogg's number.
Winnie marks the number 45 units to the right of Lizzie's number.
What number does Winnie mark?

100. _____

101. ★ In the statement below, each shape stands for a digit, and different shapes stand for different digits.

$$\triangle\square\bigcirc < 3\square\bigcirc < 3\bigcirc\square < 3\triangle\triangle$$

What digit does each shape stand for?

101. $\triangle$ = _____

$\bigcirc$ = _____

$\square$ = _____

PRACTICE | Answer each question below.

102. Alba fills every blank below with the **same digit** to make
★ the comparison true. What digit does Alba use?

102. _____

$$\square 7 \square < \square\square 0 < 9\square 0$$

103. Tia's favorite number is between 777 and 999.
★ The ones digit is less than the tens digit.
The hundreds digit is less than the ones digit.
What is Tia's favorite number?

103. _____

104. Winnie arranges seven whole numbers in order, then covers the five numbers
★ in the middle with shape cards. List all of the numbers that could be hidden
under the ★ card.

$$15 < \boxed{\hexagon} < \boxed{\triangle} < \boxed{\bigstar} < \boxed{\blacksquare} < \boxed{\blacklozenge} < 25$$

104. _____

PRACTICE | Answer each question below.

105. Fill the blanks below using only the digits 1, 3, and 6. Some of these digits
★ may be used more than once.

☐☐ is halfway between ☐☐ and ☐☐ .

106. Each incorrect statement below can be corrected by swapping two of the
★ digits. Swap **two** digits in each statement to make it correct.

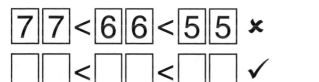

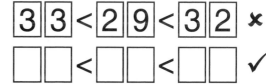

$\boxed{7}\boxed{7}<\boxed{6}\boxed{6}<\boxed{5}\boxed{5}$ ✗
☐☐<☐☐<☐☐ ✓

$\boxed{3}\boxed{3}<\boxed{2}\boxed{9}<\boxed{3}\boxed{2}$ ✗
☐☐<☐☐<☐☐ ✓

$\boxed{1}\boxed{2}<\boxed{1}\boxed{2}<\boxed{1}\boxed{2}$ ✗
☐☐<☐☐<☐☐ ✓

$\boxed{2}\boxed{2}<\boxed{5}\boxed{5}<\boxed{2}\boxed{2}$ ✗
☐☐<☐☐<☐☐ ✓

CHAPTER 3
Addition

Use this Practice book with Guide 2A from BeastAcademy.com.

Recommended Sequence:

Book	Pages:
Guide:	70-75
Practice:	65-69
Guide:	76-87
Practice:	70-85
Guide:	88-93
Practice:	86-93

You may also read the entire chapter in the Guide before beginning the Practice chapter.

EXAMPLE | Add 25+34.

25 is 2 tens and 5 ones.
34 is 3 tens and 4 ones.

So, adding 25+34 gives
2+3 = 5 tens and 5+4 = 9 ones.

5 tens is 50, so 5 tens and
9 ones is 50+9 = **59**.

The result of addition is called a **sum**.

To find a sum, we can add the place values separately.

$$\underset{\text{tens}}{2} + \underset{\text{tens}}{3} = \underset{\text{tens}}{5}$$

$$25+34 = 50+9 = 59$$

$$\underset{\text{ones}}{5} + \underset{\text{ones}}{4} = \underset{\text{ones}}{9}$$

PRACTICE | Fill the empty blanks to find each sum below.

1. 32+56 = __80__ + __8__ = _____

2. 13+51 = _____ + __4__ = _____

3. 62+34 = __90__ + ____ = _____

4. 54+21 = _____ + ____ = _____

5. 217+461 = __600__ + _____ + ____ = _____

6. 47+32 = _____

7. 61+25 = _____

8. 136+53 = _____

9. 64+315 = _____

10. 250+37 = _____

11. 270+604 = _____

12. 316+260 = _____

ADDITION
Sums

We can only have one digit in each place value.

So, sometimes we need to *regroup*.

EXAMPLE | Add 73+84.

73 is 7 tens and 3 ones.
84 is 8 tens and 4 ones.

So, adding 73+84 gives
7+8 = 15 tens and 3+4 = 7 ones.

15 tens is the same as 1 hundred and 5 tens, or 150.

10 ones make 1 ten, and 10 tens make 1 hundred.

$$\underset{tens}{7} + \underset{tens}{8} = \underset{tens}{15}$$

$$73+84 = 150+7 = 157$$

$$\underset{ones}{3} + \underset{ones}{4} = \underset{ones}{7}$$

So, 73+84 = 150+7 = 157.

PRACTICE | Fill the empty blanks to find each sum below.

13. 43+83 = __120__ + __6__ = ____

14. 72+46 = ____ + __8__ = ____

15. 54+19 = __60__ + ____ = ____

16. 36+28 = __50__ + ____ = ____

17. 71+83 = ____ + __4__ = ____

18. 59+39 = __80__ + ____ = ____

19. 52+81 = ____ + ____ = ____

20. 67+28 = ____ + ____ = ____

21. 38+83 = ____ + ____ = ____

22. 85+65 = ____ + ____ = ____

PRACTICE | Fill the empty blanks to find each sum below.

23. $223+195 =$ ___300___ + ___110___ + __8__ = _____

24. $528+345 =$ ___800___ + __60__ + _____ = _____

25. $393+461 =$ ___700___ + _____ + __4__ = _____

26. $315+167 =$ _____ + _____ + _____ = _____

27. $248+590 =$ _____ + _____ + _____ = _____

28. $273+64 =$ _____ + _____ + _____ = _____

29. $883+109 =$ _____ + _____ + _____ = _____

30. $555+346 =$ _____ + _____ + _____ = _____

Fill the empty boxes so that the sums from left-to-right and from top-to-bottom are all correct.

EXAMPLE | Complete the Cross-Number puzzle below.

Across:

$20 + 34 = 54$.

$21 + 23 = 44$.

$41 + 57 = 98$.

Down:

$20 + 21 = 41$.

$34 + 23 = 57$.

$54 + 44 = 98$.

The completed Cross-Number puzzle is shown on the right.

20	+	34	=	54
+		+		+
21	+	23	=	44
=		=		=
41	+	57	=	98

Find more Cross-Number puzzles at BeastAcademy.com.

PRACTICE | Complete each Cross-Number puzzle below.

31.

10	+	11	=	
+		+		+
15	+	12	=	
=		=		=
	+		=	

32.

70	+	20	=	
+		+		+
110	+	40	=	
=		=		=
	+		=	

PRACTICE | Complete each Cross-Number puzzle below.

33.

40	+	36	=	
+		+		+
37	+	20	=	
=		=		=
	+		=	

34.

22	+	34	=	
+		+		+
45	+	41	=	
=		=		=
	+		=	

35.

127	+	60	=	
+		+		+
40	+	172	=	
=		=		=
	+		=	

36.

32	+		=	62
+		+		+
41	+	44	=	
=		=		=
	+		=	

37. ★

120	+	70	=	
+		+		+
30	+		=	
=		=		=
	+		=	260

38. ★

191	+		=	
+		+		+
	+	212	=	335
=		=		=
	+		=	827

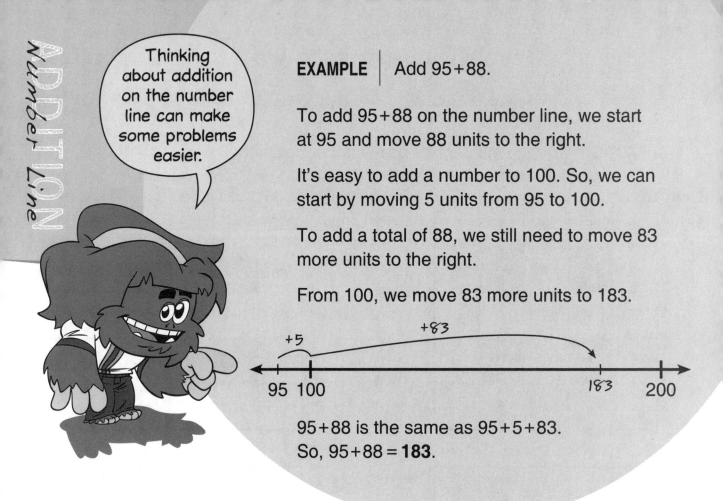

Thinking about addition on the number line can make some problems easier.

EXAMPLE | Add 95+88.

To add 95+88 on the number line, we start at 95 and move 88 units to the right.

It's easy to add a number to 100. So, we can start by moving 5 units from 95 to 100.

To add a total of 88, we still need to move 83 more units to the right.

From 100, we move 83 more units to 183.

95+88 is the same as 95+5+83.
So, 95+88 = **183**.

PRACTICE | Fill the empty blanks on the number lines below to find each sum.

39. 196+36

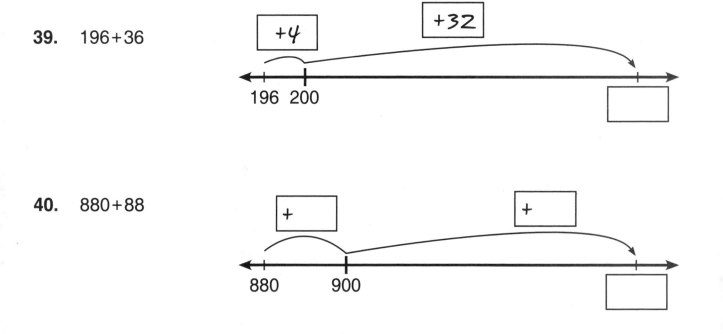

40. 880+88

PRACTICE | Fill the empty blanks on the number lines below to find each sum.

41. 350+57

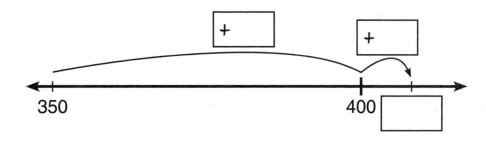

42. 175+29

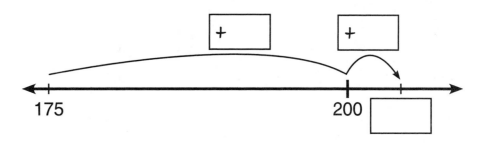

43. 391+39

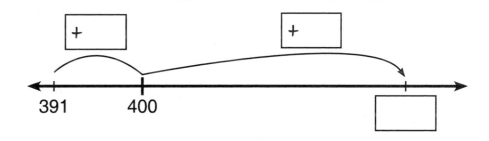

44. 193+22 = _____ **45.** 496+56 = _____ **46.** 598+223 = _____

EXAMPLE

Igor's basket contains 49 tennis balls. Marjorie's basket contains 45 tennis balls. How many tennis balls are in both baskets?

Moving tennis balls from one basket to the other does not change the total number of balls.

To make the addition easier, we take one ball from Marjorie's basket and put it in Igor's. So, instead of adding 49+45, we add 50+44.

$$49 / + \backslash 45 = 50 / + \backslash 44$$

So, 49+45 = 50+44 = **94**.

PRACTICE | Fill the blanks to find each sum below.

47. $39+26 = \boxed{40} + \boxed{} = \boxed{}$

48. $290+86 = \boxed{300} + \boxed{} = \boxed{}$

49. $52+128 = \boxed{} + \boxed{130} = \boxed{}$

50. $78+96 = \boxed{} + \boxed{100} = \boxed{}$

51. $393+128 = \boxed{} + \boxed{} = \boxed{}$

52. $595+237 = \boxed{} + \boxed{} = \boxed{}$

PRACTICE | Answer each question below.

53. Marie has two pitchers, with 96 ounces of water in each pitcher. She pours 4 ounces from one pitcher into the other. How many total ounces of water are in the two pitchers?

53. _____

54. A parking garage has a lower lot and an upper lot. There are 277 cars in the lower lot, and 203 cars in the upper lot. After three cars drive from the upper lot to the lower lot, how many cars are there in the whole garage?

54. _____

55. A bookcase has two shelves. There are 93 books on the top shelf and 78 books on the bottom shelf. How many books could you **move** from the bottom shelf to the top shelf to make it easier to find the total number of books?

55. _____

56. Sander has 98 peas on his plate. Ada and Parker each have 101 peas on their plates. What is the total number of peas on all three plates?

56. _____

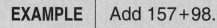

When adding, sometimes it's useful to add a little extra, then take the extra away.

EXAMPLE | Add 157+98.

Adding 98 to a number is the same as adding 100, then taking away 2.

$$157+98 = 157+100-2.$$

We have $157+100 = 257$, and $257-2 = 255$.

So, $157+98 = \mathbf{255}$.

PRACTICE | Fill the blanks to answer each question below.

57. Adding 38 to a number is the same as adding 40, then taking away _____.

58. Adding 19 to a number is the same as adding _____, then taking away 1.

59. Adding _____ to a number is the same as adding 100, then taking away 7.

60. $133+28 =$ | 133 | + | 30 | − | ☐ | = | ☐

61. $216+91 =$ | 216 | + | ☐ | − | 9 | = | ☐

62. $256+95 =$ | 256 | + | ☐ | − | ☐ | = | ☐

EXAMPLE | Add 19+19+19+19.

Since 19 is 1 less than 20, the sum of four 19's is 4 less than the sum of four 20's.

So, 19+19+19+19 = 20+20+20+20−4.

Adding 20+20+20+20 gives 80. Then, 80−4 gives **76**.

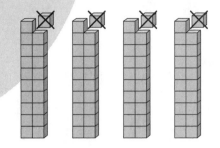

PRACTICE | Fill the blanks to answer each question below.

63. 27+27+27 = $\boxed{30}$ + $\boxed{30}$ + $\boxed{30}$ − $\boxed{}$ = $\boxed{}$

64. 99+99+99+99+99 = $\boxed{500}$ − $\boxed{}$ = $\boxed{}$

65. 298+299+300 = $\boxed{}$ − $\boxed{3}$ = $\boxed{}$

66. 25+24+23+22 = $\boxed{100}$ − $\boxed{}$ = $\boxed{}$

67. 9+19+29+39 = $\boxed{}$ − $\boxed{4}$ = $\boxed{}$

68. 190+190+190+190 = $\boxed{}$ − $\boxed{}$ = $\boxed{}$

There's more than one way to find a sum.

PRACTICE | Find each sum below.

69. $328 + 96 = \underline{\hspace{1.5cm}}$

70. $589 + 96 = \underline{\hspace{1.5cm}}$

71. $702 + 96 = \underline{\hspace{1.5cm}}$

72. $538 + 140 = \underline{\hspace{1.5cm}}$

73. $190 + 140 = \underline{\hspace{1.5cm}}$

74. $295 + 140 = \underline{\hspace{1.5cm}}$

75. $713 + 65 = \underline{\hspace{1.5cm}}$

76. $135 + 65 = \underline{\hspace{1.5cm}}$

77. $470 + 65 = \underline{\hspace{1.5cm}}$

78. $339 + 111 = \underline{\hspace{1.5cm}}$

79. $299 + 111 = \underline{\hspace{1.5cm}}$

80. $77 + 111 = \underline{\hspace{1.5cm}}$

81. $75 + 426 = \underline{\hspace{1.5cm}}$

82. $57 + 75 = \underline{\hspace{1.5cm}}$

83. $680 + 75 = \underline{\hspace{1.5cm}}$

In a Sum Pyramid, the number in each block is the sum of the two numbers below it.

EXAMPLE | Complete the Sum Pyramid below.

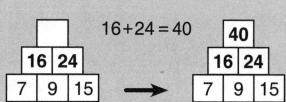

$7+9=16$
$9+15=24$
$16+24=40$

PRACTICE | Complete each Sum Pyramid below.

84.

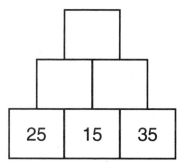

85.

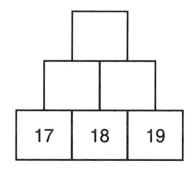

86.

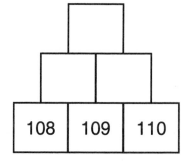

87.

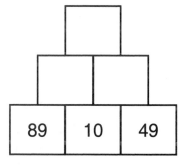

88.

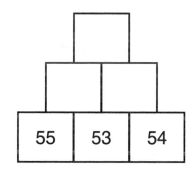

89.

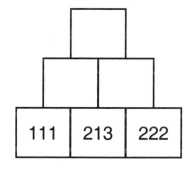

90.
★

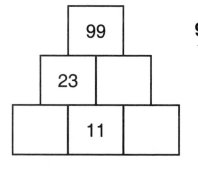

91.
★

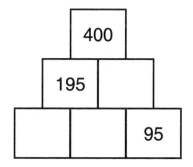

92.
★
★

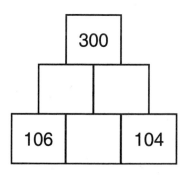

EXAMPLE | Double 95.

To double 95, we add 95 + 95.

We can add by place value.

90 + 90 = 180 and 5 + 5 = 10.
So, 95 + 95 = 180 + 10 = **190**.

— *or* —

95 is 5 less than 100. So, 95 + 95 is
5 + 5 = 10 less than 100 + 100 = 200.

So, 95 + 95 = 200 − 10 = **190**.

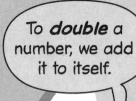

To *double* a
number, we add
it to itself.

PRACTICE | Solve each problem below.

93. 34 + 34 = _____

94. 38 + 38 = _____

95. 70 + 70 = _____

96. 234 + 234 = _____

97. 382 + 382 = _____

98. 390 + 390 = _____

99. What number can be doubled to give the same
result as 13 + 13 + 13 + 13?

99. _____

100. What number can be doubled to give the same
result as 248 + 252?

100. _____

PRACTICE | Double each number to get the next number in the rows below.

101. | 3 | 6 | 12 | 24 | | | | | 768 |

102. | 2 | 4 | 8 | | | | | 256 | |

103. | 5 | 10 | | | | 160 | | |

104. | 7 | | | | 112 | | | |

105. | 11 | | | | | | | |

106. ★ Ralph doubles a number, then doubles the result and gets 92. What number did Ralph start with?

106. _____

107. ★ Winnie doubles 88. Grogg adds two **2-digit** numbers and gets the same sum as Winnie. What is the smallest number that could be part of Grogg's sum?

107. _____

ADDITION

EXAMPLE | Find the ones digit of 75+64+53.

We can add by place value. 70+60+50 = 180, and 5+4+3 = 12. So, 75+64+53 = 180+12 = 192.

The ones digit of 192 is **2**.

— *or* —

Adding tens to a number doesn't change its ones digit.

So, to find the ones digit of 75+64+53, we can ignore the tens and just add the ones: 5+4+3 = 12.

The ones digit of 75+64+53 is the same as the ones digit of 5+4+3 = 12, which is **2**.

The ones digit of a sum can be found without finding the other digits.

PRACTICE | Answer each question below.

108. Ralph adds 12+22+32+42+52+62+72+82+92. What is the **ones digit** of Ralph's result?

108. _____

109. Circle the two numbers below that have a sum with ones digit 4.

243   244   245   246   247   248

110. What is the smallest number of 97's that can be added to give a result with ones digit 8?

110. _____

111. ★ Penny adds 111 copies of 111. What is the ones digit of her result?

111. _____

Fruit	Price (¢)
Strawberry	9
Banana	19
Orange	29
Apple	39
Dragonfruit	49
Mango	99

This table shows the cost in cents of fruits at the Beast Market.

PRACTICE | Use the table above to answer each question below.

112. Find the cost of each of these purchases.

2 Bananas: _____ cents   5 Strawberries: _____ cents

3 Mangos: _____ cents   4 Dragonfruits: _____ cents

113. Circle the only amount below that could be the total cost of 6 fruits.

233 cents   234 cents   235 cents   236 cents

114. Winnie buys a bag of fruit for 186 cents. No two fruits in the bag are the same. How many fruits are in Winnie's bag?

114. _____

115. Alex buys a bag of fruit for 108 cents. All of the fruits in Alex's bag are the same. What fruits did Alex buy?

115. _____

116. What is the largest number of fruits that can be bought for exactly 127 cents?

116. _____

EXAMPLE | Fill the blank: 35 + □ = 78.

78 has 4 more tens and 3 more ones than 35.

So, to get from 35 to 78, we add
4 tens and 3 ones, which is **43**.

Check: 35 + 43 = 78.

We can find
the missing
number in a
sum.

EXAMPLE | Fill the blank: 29 + □ = 86.

86 has 6 more tens than 29. But, adding 6 tens to 29
gives 89. Since this is too large, we try 5 tens.

Adding 5 tens to 29 gives 79.
Then, adding 7 ones gives 86.

So, to get from 29 to 86, we add
5 tens and 7 ones, which is **57**.

Check: 29 + 57 = 86.

PRACTICE | Fill the blank in each sum below.

117. 34 + □ = 88

118. 16 + □ = 58

119. 30 + □ = 91

120. 24 + □ = 70

121. 56 + □ = 93

122. 154 + □ = 182

123. 314 + □ = 478

124. 160 + □ = 240

125. 333 + □ = 518

After the first two numbers in a **Sum String**, each number is the sum of the two numbers to its left. For example, the Sum String below starts with 3 and 10. So, the third number is 3+10 = 13, the fourth number is 10+13 = 23, the fifth number is 13+23 = 36, and so on.

EXAMPLE | Find the missing numbers in the Sum String below.

To the right of 27 and 43 is 27+43 = 70.

Next is 43+70 = 113, followed by 70+113 = 183.

Since $\boxed{11}$+16 = 27, the first number in this Sum String is 11.

PRACTICE | Fill in the missing numbers in each Sum String below.

126.
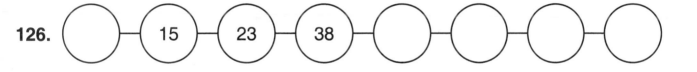

127.
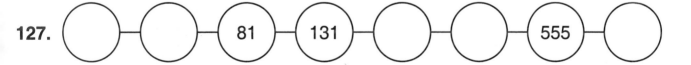

128. ★
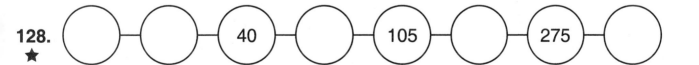

129. ★
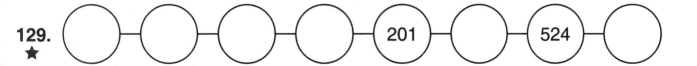

It's helpful to recognize pairs of numbers that combine to make 10's or 100's.

In the problems below, circle pairs of numbers so that every number is in one pair. The circled pairs must be in shapes that touch.

Below, 4 pairs of numbers are circled so that every pair has a sum of 100.

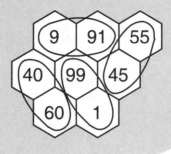

PRACTICE | Solve each problem below.

130. Circle eight pairs of numbers so that every pair has a sum of 100.

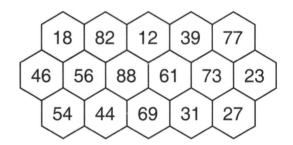

131. Circle eight pairs of numbers so that every pair has a sum that ends in 0.

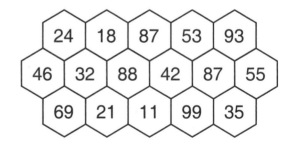

132. ★ Circle eight pairs of numbers so that every pair has a sum that ends in 0.

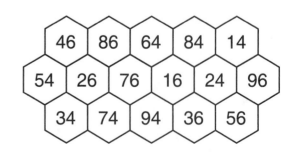

In a 100-Triangle, the numbers in the three circles on each side add up to 100.

EXAMPLE | Complete the 100-Triangle below.

Left: $61 + 15 + 24 = 100$.

Right: $24 + 63 + 13 = 100$.

Bottom: $61 + 26 + 13 = 100$.

So, we have:

PRACTICE | Complete each 100-Triangle below.

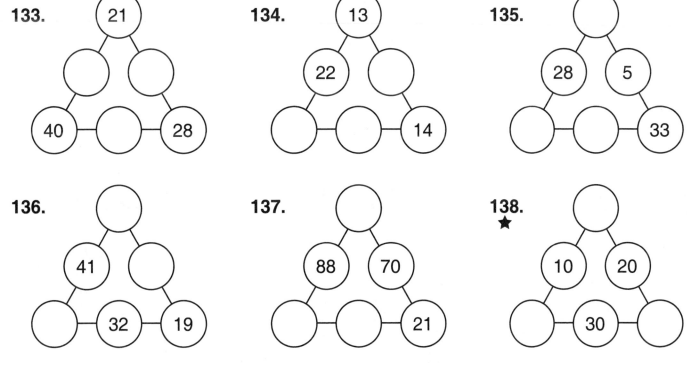

Find more 100-Triangle puzzles at BeastAcademy.com.

You can add numbers in any order!

When adding two numbers, order does not matter. For example, 5+13 and 13+5 both equal 18.

$$5+13$$
$$= 13+5$$
$$= 18$$

When adding more than two numbers, it doesn't matter which two we add first.
To add 18+25+75, we can start by adding 18+25. But, it is easier to start by adding 25+75 = 100. Then, 18+100 = 118.

$$18+25+75$$
$$= 18+100$$
$$= 118$$

PRACTICE | Fill the blanks to answer each question below.

139. $38+91+9$

$= 38 + \boxed{}$

$= \boxed{}$

140. $15+27+15$

$= 27 + \boxed{}$

$= \boxed{}$

141. $16+33+14+7$

$= \boxed{} + \boxed{}$

$= \boxed{}$

142. $39+12+38$

$= \boxed{} + \boxed{}$

$= \boxed{}$

143. $49+36+51$

$= \boxed{} + \boxed{}$

$= \boxed{}$

144. $38+38+32+32$

$= \boxed{} + \boxed{}$

$= \boxed{}$

PRACTICE | Fill the blanks to answer each question below.

145. $44+23+77+18+32$

$= \boxed{} + \boxed{} + \boxed{}$

$= \boxed{}$

146. $58+20+42+39+80$

$= \boxed{} + \boxed{} + \boxed{}$

$= \boxed{}$

147. $79+21+85+15+56$

$= \boxed{} + \boxed{} + \boxed{}$

$= \boxed{}$

148. $119+226+431+74$

$= \boxed{} + \boxed{}$

$= \boxed{}$

149. What is the sum of five 16's and five 34's?

149. _____

150. What is the sum $15+17+19+21+23+25$?

150. _____

151. ★ What is the sum of every whole number from 1 to 19?

151. _____

ADDITION

EXAMPLE | In the row of digits below, circle two 3-digit numbers whose sum is 900.

5 4 5 5 5 4 4 4 5 4

Since 900 has ones digit 0, two numbers that sum to 900 must have ones digits whose sum ends in 0. The only way to get a sum ending in 0 using two digits from the row is 5+5 = 10.

So, both numbers we circle must have ones digit 5.

We find 455+445 = 900.

5 4 5 5 5 4 4 4 5 4

The numbers you circle in these problems cannot overlap.

PRACTICE | Circle two numbers in each row of digits that have the given sum. The numbers you circle cannot overlap.

152. Circle two 2-digit numbers whose sum is 70.

1 1 2 2 4 4 8 8

153. Circle two 2-digit numbers whose sum is 100.

2 3 3 4 4 5 5 6

154. Circle two 3-digit numbers whose sum is 975.

9 8 7 6 5 4 3 2 1

155. Circle two 3-digit numbers whose sum is 789.

1 2 3 3 4 5 5 6 7 7 8 9 9

PRACTICE | Circle two numbers in each row of digits that have the given sum. The numbers you circle cannot overlap.

156. Circle two 3-digit numbers whose sum is 432.

1 2 2 1 1 2 2 1

157. Circle two 3-digit numbers whose sum is 710.

7 3 3 3 7 3 3 3 7

158. Circle two 3-digit numbers whose sum is 796.

1 2 3 4 5 1 2 3 4 5

159. Circle two 3-digit numbers whose sum is 558.

1 4 4 1 4 1 1 1 4 4 4

160. Circle two 3-digit numbers whose sum is 850.

1 1 1 7 7 7 3 3 3 5 5 5

161. Circle two 3-digit numbers whose sum is 800.

5 5 5 7 7 7 8 8 8 2 2 2 3 3 3

The numbers in each blob must be in squares that share a side.

Every number must be in exactly one blob, and the blobs cannot cross each other.

EXAMPLE

Circle blobs of two or more numbers in the grid below so that the sum of the numbers in each blob is 70.

7	21	42
28	21	28
14	21	28

The sums in the three blobs are
$14+28+7+21 = 70$,
$21+21+28 = 70$, and
$42+28 = 70$.

PRACTICE | Circle blobs of two or more numbers in each grid below. The sum of the numbers in each blob must equal the target.

---

162. Target: 50

33	26	24
17	13	23
23	27	14

163. Target: 39

7	10	19
2	10	12
30	14	13

164. Target: 88

11	44	55
11	22	33
11	66	11

---

165. Target: 15

5	4	3
6	6	2
7	8	4

166. Target: 100

73	10	17
27	93	7
23	13	37

167. Target: 25

18	2	3
2	1	6
21	3	19

PRACTICE | Circle blobs of two or more numbers in each grid below. The sum of the numbers in each blob must equal the target.

168. Target: 120

72	58	30
20	28	22
38	82	10

169. Target: 600

224	324	276
276	176	424
100	524	76

170. Target: 101

11	14	11
47	76	43
23	31	47

171. Target: 50

10	20	30	29
10	10	7	14
20	13	17	20
11	12	13	14

172. Target: 999

600	300	54	45
200	54	45	100
700	500	54	45
400	54	45	800

173. Target: 400

100	96	108	97
101	100	104	92
98	94	100	103
99	102	106	100

174. Target: 99

22	11	22	22
33	10	11	10
22	20	11	20
44	13	13	13

PRACTICE | Answer each question below.

175. Adam adds nine 79's. What is the **ones digit** of the result?

175. _____

176. What is $10+11+20+22+30+33+40$?

176. _____

177. What number can you double to equal the sum
★ of five 222's?

177. _____

178. There are 3 ways to add two 1-digit numbers to get
★ a sum of 16:

$$9+7, \ 8+8, \text{ and } 7+9.$$

How many ways are there to add two 3-digit numbers
to get a sum of 211?

178. _____

PRACTICE | Answer each question below.

179. Grogg adds three of the numbers below and gets a sum with ones digit 3. What is Grogg's sum?
★

$$45 \quad 56 \quad 67 \quad 78 \quad 89$$

179. _____

180. Lizzie writes a **different** digit in each blank below to create a sum of three 3-digit numbers. What is the smallest possible sum of the three numbers?
★

$$\underline{}\,\underline{}\,\underline{} + \underline{}\,\underline{}\,\underline{} + \underline{}\,\underline{}\,\underline{}$$

180. _____

181. Winnie uses the numbers 18, 21, 25, 36, 40, and 43 to make three pairs of numbers that have the same sum. What is the sum of each pair of numbers?
★

$$\boxed{}+\boxed{} = \boxed{}+\boxed{} = \boxed{}+\boxed{}$$

181. _____

182. Alex splits the numbers 10, 14, 21, 22, 30, and 53 into two groups that have the same sum. What is the sum of each group of numbers?
★

182. _____

HINTS
For Selected Problems

Below are hints to every problem marked with a ★.
Work on the problems for a while before looking at the hints.
The hint numbers match the problem numbers.

9. After replacing ten •'s with an **X**, how many **X**'s are there?

22. How else can we write **XXXXXX**?

46. Where can the 7 be placed?

47. What is the ones digit: 5, 7, or 9?

55. What if the tens digits are all 1's? 2's? 3's? 4's?

56. Why can't we circle 773?

73. 3 hundreds is how many tens?

74. What do we get when we take away 12 ones from 222?

99. Adding 1 hundred to 792 gives 892. We must still add 1 ten. What numbers fill the following blanks?

892 is ___ tens and ___ ones.

118. What numbers fill the following blanks?

424 is ___ tens and ___ ones.

122. What is Alex's number?

133. What two numbers can connect 182 to 83?

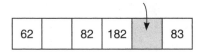

Which one must we use?

134. What two numbers can connect 91 to 100?

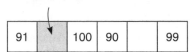

Which one must we use?

150. What two numbers go in the shaded squares below?

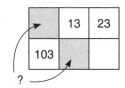

Which number goes in which square?

151. What number goes in the shaded square below?

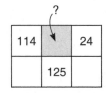

152. What are the different ways we can connect 146 to 66?

174. A 3-digit number is 100 or more. What does that tell us?

175. What numbers fill the following blanks?

358 is ___ tens and ___ ones.

176. What numbers fill the following blanks?

77 ones is ___ tens and ___ ones.

177. What is the hundreds digit of the number?

178. How many times will 5 appear in the ones place? The tens place?

11. There are six different 3-digit numbers that can be written using 2, 4, and 6 once each. Can you write them all?

21. The number halfway between 32 and 76 is the same as the number halfway between 32+10 and 76−10.

22. The number halfway between 534 and 584 is the same as the number halfway between 534+10 and 584−10.

25. The number halfway between 43 and 89 is the same as the number halfway between 43+10 and 89−10.

26. What number is halfway between 40 and 90?

27. What number is halfway between 350 and 410?

29. Before Player X's final move, Player O has 5 points and Player X has 3 points.

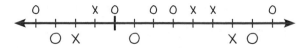

What tick mark can Player X label to take as many points away from Player O as possible?

30. Before Player X's final move, Player O has 6 points and Player X has 2 points.

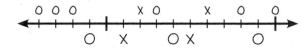

What tick mark can Player X label to take as many points away from Player O as possible?

31. Before Player X's final move, Player O has 7 points and Player X has 4 points.

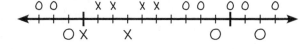

32. Before Player X's final move, Player O has 10 points and Player X has 5 points.

43. When completed, 3 will not be the smallest number in this puzzle!

44. How can we connect 401 to 405 without blocking all of the paths from 398 to 401?

45. How can we connect 15 to 19 without blocking all of the paths from 19 to 21?

46. What are the different ways we can connect 901 to 905?

48. Where must 103 go? Then, how many empty hexagons are needed to connect 104 to 113?

49. Where must 67 go? Then, how many empty hexagons are needed to connect 68 to 78?

50. How can we connect 49 to 53?

64. What digit must be placed in the blank in 46>4☐?

85. Be organized! What numbers can you write with hundreds digit 1? Hundreds digit 2? Hundreds digit 3?

86. There is 1 missing two-digit number, and 3 missing three-digit numbers.

87. There are 3 missing two-digit numbers, and 1 missing three-digit number.

95. What numbers must go in the tens digit blanks?

96. What is the smallest number we can make with the given digits? The largest number?

97. What is a good number to count in by? Think big!

98. How should we label the tick marks between 15 and 60?

99. Draw a number line!

100. Draw a number line!

101. What does ▲■◯<3■◯ tell us about the ▲?

102. Why can't Alba use 1? Why can't Alba use 9?

103. What could be the hundreds digit of Tia's number?

104. What is the smallest number that could be under ▲? What is the largest number that could be under ■?

105. What are the tens digits of the three numbers?

106. If you're stuck, try something! Even if it's wrong, it may help you make a better guess.

CHAPTER 3

Addition   64–93

37. What squares can you fill in first?

38. What number goes in the middle-left square?

90. What number do we add to 23 to get 99?

91. What number do we add to 195 to get 400?

92. Take a guess for what number belongs in the empty bottom-middle block. Did your guess work? If not, can you make a better guess? Keep trying!

106. Work backwards! What number do we double to get 92?

107. What is the *largest* number that could be part of Grogg's sum?

111. Can you ignore any of the digits in 111?

114. If Winnie bought 1 fruit, what would be the ones digit of the cost? 2 fruits? 3 fruits?

115. What is the ones digit of the cost of 1 fruit? 2 fruits? 3 fruits? Do you notice a pattern? What happens when there are 11 fruits?

116. What is the ones digit of the cost of 1 fruit? 2 fruits? 3 fruits? Do you notice a pattern? What happens when there are 11 fruits?

128. Start by finding the number you add to 40 to get 105.

129. Start by finding the number you add to 201 to get 524.

132. What number must be paired with the 46 in the top-left?

138. Guess what number belongs in the top circle. Did your guess work? If not, can you make a better guess? Keep trying!

151. What numbers pair nicely to make this sum easier?

171. What numbers must be in the same blob as 29?

172. What numbers must be in the same blob as 800?

173. All of the numbers are equal to or close to 100. We know 100+100+100+100=400. How does this help?

174. The ones digit of the sum of every blob is 9. What numbers must be in the same blob as the 13 in the bottom-right corner?

177. How could you split five 222's into two equal groups?

178. What is the smallest three-digit number that could be in the sum? What is the largest?

179. Do we need to consider the tens digits of any of the numbers?

180. What digits should we use to fill the blanks in the hundreds places?

181. Why can't we pair 18 with 21? What number must be paired with 18?

182. Can you find the sum of each group *without* splitting the numbers into two equal groups?

SOLUTIONS
Chapters 1-3

Pirate Numbers   7 – 11

1. We draw an **X** for each stack of ten coins, and a • for each extra coin. There are **X X X** • • • • • coins.

2. We draw an **X** for each stack of ten coins, and a • for each extra coin. There are **X X X X** • • • • • • • • coins.

3. We draw a **Ϲ** for each group of one hundred coins, an **X** for each stack of ten coins, and a • for each extra coin. There are **Ϲ X X X X** • • • coins.

4. We replace ten •'s with one **X**.

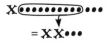

= **X X** • • •

5. We replace ten **X**'s with one **Ϲ**.

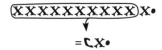

= **Ϲ X** •

6. We replace ten **X**'s with one **Ϲ**.

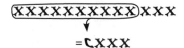

= **Ϲ X X X**

7. We replace ten •'s with one **X**.

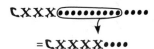

= **Ϲ X X X X** • • • •

8. We replace ten **X**'s with one **Ϲ**, and ten •'s with one **X**.

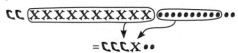

= **Ϲ Ϲ Ϲ X** • •

9. We replace each group of ten •'s with one **X**.

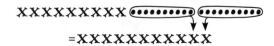

= **X X X X X X X X X X X X**

Then, we replace ten **X**'s with one **Ϲ**.

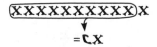

= **Ϲ X**

10. There are one **X** and eight •'s in all.

X • • • • • • • •

11. There are eight **X**'s and six •'s in all.

X X X X X X X X • • • • • •

12. There are five **X**'s and seven •'s in all.

X X X X X • • • • • • •

13. There are six **X**'s and twelve •'s in all. We replace ten •'s with one **X**.

= **X X X X X X X** • •

14. There are ten **X**'s and four •'s in all. We replace ten **X**'s with one **Ϲ**.

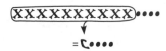

= **Ϲ** • • • •

15. There are one **Ϲ** and eleven **X**'s in all. We replace ten **X**'s with one **Ϲ**.

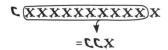

= **Ϲ Ϲ X**

16. There are nine **X**'s and ten •'s in all. We replace ten •'s with one **X**.

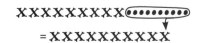

= **X X X X X X X X X X**

Then, we replace ten **X**'s with one **Ϲ**.

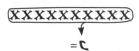

= **Ϲ**

17. Captain Kraken has **X X X X X** + **X X X X X X X** coins. There are twelve **X**'s in all.

We replace ten **X**'s with one **Ϲ**.

= **Ϲ X X**

So, Captain Kraken now has **Ϲ X X** coins.

18. Barnacle Barney has buried **Ϲ X** • • • • • • + **X X** • • • • • • • coins. There are one **Ϲ**, three **X**'s, and thirteen •'s in all.

We replace ten •'s with one **X**.

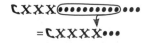

= **Ϲ X X X X** • • •

So, Barnacle Barney has buried **Ϲ X X X X** • • • coins.

19. Captain Kraken starts with **X X X** • • • coins, and then takes away **X X** • coins. Taking away two **X**'s and one • from **X X X** • • • leaves **X** • •.

So, Captain Kraken has **X** • • coins left.

20. We split the **X**'s and the •'s into two equal groups. Splitting **X X X X** coins gives two groups of **X X** coins. Splitting • • • • • • coins gives two groups of • • • coins.

So, we can make two equal piles of **X X** • • • coins.

21. There are three bags, with $\mathbf{XX}\bullet\bullet\bullet\bullet$ coins in each bag. So, there are $\mathbf{XX}\bullet\bullet\bullet\bullet + \mathbf{XX}\bullet\bullet\bullet\bullet + \mathbf{XX}\bullet\bullet\bullet\bullet$ coins all together.

This gives six $\mathbf{X}$'s and twelve $\bullet$'s.
We replace ten $\bullet$'s with one $\mathbf{X}$.

So, there are $\mathbf{XXXXXXX}\bullet\bullet$ coins in the treasure chest.

22. To take away $\mathbf{XX}\bullet\bullet\bullet\bullet$ coins from $\mathbf{XXXXXX}$ coins, we start by writing one of the $\mathbf{X}$'s in $\mathbf{XXXXXX}$ as ten $\bullet$'s.

$$\mathbf{XXXXXX} = \mathbf{XXXXX}\bullet\bullet\bullet\bullet\bullet\bullet\bullet\bullet\bullet\bullet$$

Taking away two $\mathbf{X}$'s and four $\bullet$'s from $\mathbf{XXXXX}\bullet\bullet\bullet\bullet\bullet\bullet\bullet\bullet\bullet\bullet$ leaves $\mathbf{XXX}\bullet\bullet\bullet\bullet\bullet\bullet$.

So, there are $\mathbf{XXX}\bullet\bullet\bullet\bullet\bullet\bullet$ coins left in the chest.

PLACE VALUE
Digits   12–15

23. The $\mathbf{X}$ stands for 10 $\bullet$'s, and 3 more $\bullet$'s makes 13 $\bullet$'s. So, $\mathbf{X}\bullet\bullet\bullet = \mathbf{13}$.

24. The $\mathbf{X}$ stands for 10 $\bullet$'s, and 8 more $\bullet$'s makes 18 $\bullet$'s. So, $\mathbf{X}\bullet\bullet\bullet\bullet\bullet\bullet\bullet\bullet = \mathbf{18}$.

25. Since each $\mathbf{X}$ stands for 10 $\bullet$'s, we count the $\mathbf{X}$'s by tens.

Then, we have 4 more $\bullet$'s. 4 more than 20 is 24.
So, $\mathbf{XX}\bullet\bullet\bullet\bullet = \mathbf{24}$.

26. We count the $\mathbf{X}$'s by tens.

Then, we have 1 more $\bullet$. 1 more than 60 is 61.
So, $\mathbf{XXXXXX}\bullet = \mathbf{61}$.

27. We count the $\mathbf{X}$'s by tens.

Then, we have 4 more $\bullet$'s. 4 more than 50 is 54.
So, $\mathbf{XXXXX}\bullet\bullet\bullet\bullet = \mathbf{54}$.

28. We count the $\mathbf{X}$'s by tens.

10 20 30 40
X   X   X   X

Then, we have 5 more $\bullet$'s. 5 more than 40 is 45.
So, $\mathbf{XXXX}\bullet\bullet\bullet\bullet\bullet = \mathbf{45}$.

29. We count the $\mathbf{X}$'s by tens.

Then, we have 3 more $\bullet$'s. 3 more than 20 is 23.
So, $\mathbf{XX}\bullet\bullet\bullet = \mathbf{23}$.

30. We count the $\mathbf{X}$'s by tens.

So, $\mathbf{XXXXX} = \mathbf{50}$.

31. We have 1 hundred, 4 tens, and 3 ones. So, $\mathsf{C}\,\mathbf{XXXX}\bullet\bullet\bullet = \mathbf{143}$.

32. We have 1 hundred, 1 ten, and 8 ones. So, $\mathsf{C}\,\mathbf{X}\bullet\bullet\bullet\bullet\bullet\bullet\bullet\bullet = \mathbf{118}$.

33. We have 3 hundreds, 2 tens, and 1 one. So, $\mathsf{CCC}\,\mathbf{XX}\bullet = \mathbf{321}$.

34. We have 1 hundred, 2 tens, and 4 ones. So, $\mathsf{C}\,\mathbf{XX}\bullet\bullet\bullet\bullet = \mathbf{124}$.

35. We have 5 hundreds, 1 ten, and 2 ones. So, $\mathsf{CCCCC}\,\mathbf{X}\bullet\bullet = \mathbf{512}$.

36. We have 1 hundred, 0 tens, and 5 ones. So, $\mathsf{C}\,\bullet\bullet\bullet\bullet\bullet = \mathbf{105}$.

37. We have 1 hundred, 3 tens, and 0 ones. So, $\mathsf{C}\,\mathbf{XXX} = \mathbf{130}$.

38. We have 7 hundreds, 0 tens, and 0 ones. So, $\mathsf{CCCCCCC} = \mathbf{700}$.

39. Only 55, 25, and 52 have two digits.

255　(55)　(25)　522　555　(52)

40. Only 2̲09 and 2̲90 have 2 in the hundreds place.

(209)　902　92　29　(290)　920

41. Only 47̲7, 77̲4, and 47̲4 have tens digit 7.

(477)　747　(774)　447　(474)　744

42. Only 978, 897, and 987 have a hundreds digit larger than their ones digit.

(978)　798　879　(897)　789　(987)

43. We write 3 in the tens place, giving _3_. The remaining digits are 2 and 2, so we write them in the remaining place values. This gives **232**.

44. We write the 7 in the ones place, giving _ _7. The two remaining digits are 5 and 0. We cannot write 0 as the leftmost digit of a three-digit number. So, we write 5 in the hundreds place and 0 in the tens place. This gives **507**.

45. No two-digit number can have tens digit 0. So, 0 must go in the ones place, giving _0.

We can write any digit other than 0 in the tens place (1, 2, 3, 4, 5, 6, 7, 8, or 9). So, there are **9** different two-digit numbers that have 0 as a digit:

10, 20, 30, 40, 50, 60, 70, 80, and 90.

46. 7 can be the hundreds digit, tens digit, or ones digit. The other digits are both 8's. So, the three different three-digit numbers that use the digits 7, 8, and 8 are **788**, **878**, and **887**.

47. The ones digit is larger than the tens digit. So, the ones digit cannot be the smallest digit, 5.

The ones digit is smaller than the hundreds digit. So, the ones digit cannot be the largest digit, 9.

So, the ones digit is 7. Since the ones digit is larger than the tens digit, the tens digit is 5. Since the ones digit is smaller than the hundreds digit, the hundreds digit is 9.

So, the number is **957**.

Number Search 16-17

48. There is only one 3-digit number with ones digit 2 and hundreds digit 4.

2 3 4 2 (4 3 2)

49. The different 3-digit numbers in this row are 554, 545, and 455. Only 554 and 455 have tens digit 5. There is only one way to circle these two numbers without overlapping.

(5 5 4) 5 5 (4 5 5)

50. The different 3-digit numbers in this row with ones digit 9 are 999, 989, and 899. There is only one way to circle two of these numbers without overlapping.

(9 9 9)(8 9 9) 9

51. The only 3-digit numbers in this row with ones digit 0 are 110 and 100. There is only one way to circle these numbers without overlapping.

0 0 1 (1 0 0)(1 1 0)

52. The only 3-digit numbers in this row with ones digit 0 are 120, 200, and 220. There is only one way to circle these three numbers without overlapping.

0 1 (2 0 0)(2 2 0) 0 (1 2 0)

53. 314 is the only 3-digit number that appears three times in the row.

1 2 (3 1 4) 2 1 (3 1 4) 1 2 (3 1 4) 2

54. The only 3-digit numbers in this row whose tens digit is the largest digit are 576, 465, and 354.

8 7 6 (5 7 6) 5 (4 6 5) 4 (3 5 4) 3 2

55. We look at each possible tens digit.

If the tens digit of each number is 1, we can only circle three numbers without overlapping.

If the tens digit of each number is 3, we can only circle two numbers without overlapping.

The same is true if the tens digit is 4.

If the tens digit is 2, we can circle four numbers without overlapping.

4 (3 2 1)(1 2 3) 4 3 (1 2 4)(4 2 1) 3

56. The 3-digit numbers in this row are 733, 337, 373, 377, and 773. There is only way to circle four of these numbers without overlapping.

(7 3 3) 7 3 (3 7 7)(3 3 7) 3 (3 7 3)

57. There are 2 hundreds, 1 ten, and 5 ones. So, there are **215** blocks.

58. There are 2 tens, 9 ones, and 1 hundred. Ordering the place values from largest to smallest, we have 1 hundred, 2 tens, and 9 ones. So, there are **129** blocks.

59. There are **18** blocks.

60. There are 11 tens. We combine 10 tens to make 1 hundred. This gives 1 hundred, 1 ten, and 0 ones. So, there are **110** blocks.

61. There are 5 tens and 15 ones. We combine 10 ones to make 1 ten. This gives 6 tens and 5 ones. So, there are **65** blocks.

62. There are 10 tens and 4 ones. We combine 10 tens to make 1 hundred. This gives 1 hundred, 0 tens, and 4 ones. So, there are **104** blocks.

63. There are 2 hundreds, 9 tens, and 29 ones. We combine 20 ones to make 2 tens. This gives 2 hundreds, 11 tens, and 9 ones.

Then, we combine 10 tens to make 1 hundred. This gives 3 hundreds, 1 ten, and 9 ones.

So, there are **319** blocks.

64. Every place value must contain a single digit. We cannot write a number with 17 in the ones place. So, we need to regroup. We can regroup 10 ones to make 1 ten.

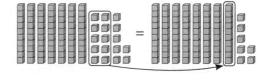

So, 7 tens and 17 ones is the same as 8 tens and 7 ones, which is **87**.

65. We cannot write a number with 28 in the ones place. So, we regroup 20 ones to make 2 tens.

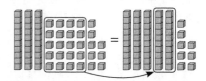

So, 3 tens and 28 ones is the same as 5 tens and 8 ones, which is **58**.

66. We cannot write a number with 11 in the ones place. So, we regroup 10 ones to make 1 ten.

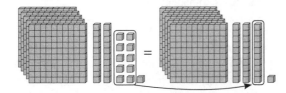

5 hundreds, 2 tens, and 11 ones is the same as 5 hundreds, 3 tens, and 1 one, which is **531**.

67. We cannot write a number with 15 in the tens place. So, we regroup 10 tens to make 1 hundred.

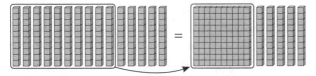

15 tens and 0 ones is the same as 1 hundred, 5 tens, and 0 ones, which is **150**.

68. We have 5 tens and 26 ones. To get 6 tens, we need 1 more ten. We regroup 10 ones to make 1 ten:

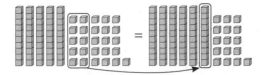

5 tens and 26 ones is the same as 6 tens and **16** ones.

69. We have 5 tens and 26 ones. To get 7 tens, we need 2 more tens. We regroup 20 ones to make 2 tens:

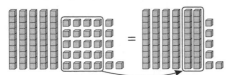

5 tens and 26 ones is the same as 7 tens and **6** ones.

70. We have 7 tens and 14 ones. To get 8 tens, we need 1 more ten. We regroup 10 ones to make 1 ten:

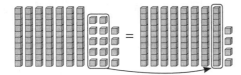

7 tens and 14 ones is the same as 8 tens and **4** ones.

71. We have 7 tens and 14 ones. To get 6 tens, we need 1 less ten. We break 1 ten into 10 ones:

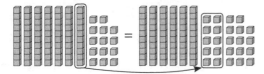

7 tens and 14 ones is the same as 6 tens and **24** ones.

72. We have 12 tens. To leave 2 tens, we regroup 10 tens to make 1 hundred. So, 12 tens is the same as **1** hundred and 2 tens.

73. We have 3 hundreds and 7 tens. To leave 0 hundreds, we break each of the hundreds into 10 tens. This gives $10+10+10=30$ more tens. So, 3 hundreds and 7 tens is the same as $30+7=$ **37** tens.

74. We have 2 hundreds, 2 tens, and 2 ones. To get 12 ones, we break 1 ten into 10 ones. This gives 2 hundreds, 1 ten, and 12 ones.

Then, to get 0 hundreds, we break each of the hundreds into 10 tens. This gives $10+10=20$ more tens for a total of $20+1=21$ tens.

So, 222 is the same as **21** tens and 12 ones.

75. We have 6 tens and **2** ones, which is 62.

To leave 5 tens, we break 1 of the 6 tens. Breaking 1 ten makes 10 ones.

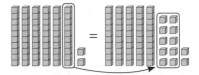

So, 62 is 5 tens and $10+2=$ **12** ones.

To leave 4 tens, we break 2 of the 6 tens. Breaking 2 tens makes 20 ones.

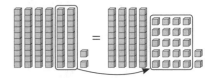

So, 62 is 4 tens and $20+2=$ **22** ones.

To leave 3 tens, we break 3 of the 6 tens. Breaking 3 tens makes 30 ones.

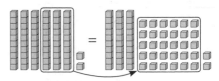

So, 62 is 3 tens and $30+2=$ **32** ones.

We fill the chart as shown.

tens	ones
6	2
5	12
4	22
3	32

76. We have **4** tens and 7 ones, which is 47.

To leave 3 tens, we break 1 of the 4 tens. Breaking 1 ten makes 10 ones.

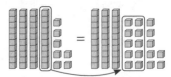

So, 47 is 3 tens and $10+7=$ **17** ones.

To get 27 ones, we break 2 tens to make 20 ones.

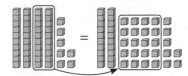

So, 47 is **2** tens and 20+7 = **27** ones.

To leave 1 ten, we break 3 of the tens.
Breaking 3 tens makes 30 ones.

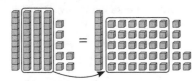

So, 47 is 1 ten and 30+7 = **37** ones.

We fill the chart as shown.

tens	ones
4	7
3	**17**
2	27
1	**37**

77. 39 is:

- 0 tens and **39** ones
- **1** ten and 29 ones
- 2 tens and **19** ones
- **3** tens and 9 ones

We fill the chart as shown.

tens	ones
0	**39**
1	29
2	**19**
3	9

78. 42 is:

- 4 tens and **2** ones
- 3 tens and **12** ones
- **2** tens and 22 ones
- 1 ten and **32** ones

We fill the chart as shown.

tens	ones
4	**2**
3	**12**
2	22
1	**32**

79. 58 is:

- **1** ten and 48 ones
- **4** tens and 18 ones
- 3 tens and **28** ones
- 5 tens and **8** ones

We fill the chart as shown.

tens	ones
1	48
4	18
3	**28**
5	**8**

80. 50 is:

- **5** tens and 0 ones
- 4 tens and **10** ones
- 3 tens and **20** ones
- 1 ten and 40 ones

We fill the chart as shown.

tens	ones
5	0
4	**10**
3	**20**
1	40

81. 91 is:

- 7 tens and **21** ones
- **1** ten and 81 ones
- **2** tens and 71 ones
- **9** tens and 1 one

We fill the chart as shown.

tens	ones
7	**21**
1	81
2	71
9	1

82. 77 is:

- **6** tens and 17 ones
- 1 ten and **67** ones
- **7** tens and 7 ones
- 0 tens and **77** ones

We fill the chart as shown.

tens	ones
6	17
1	**67**
7	7
0	**77**

83. 890 is 8 hundreds and 9 tens.
We can break 8 hundreds to make 80 tens.
So, 890 is 80+9 = **89** tens.

605 is 6 hundreds and 5 ones.
We can break 6 hundreds to make 60 tens.
So, 605 is **60** tens and 5 ones.

534 is the same as 5 hundreds and 34 ones.
We can break 5 hundreds to make 50 tens.
So, 534 is **50** tens and 34 ones.

84. We can regroup 10 tens to make 1 hundred.
So, 17 tens is 1 hundred and 7 tens, which is **170**.

We can regroup 90 tens to make 9 hundreds.
So, 90 tens is **900**.

We can regroup 40 of the 44 tens to make 4 hundreds.
That gives 4+4 = 8 hundreds and 4 tens, which is **840**.

PLACE VALUE
Ones, Tens, and Hundreds 24-27

85. Adding 1 to 134 increases its ones digit by 1.
So, 134+1 = **135**.

86. Adding 10 to 134 increases its tens digit by 1.
So, 134+10 = **144**.

87. Adding 100 to 134 increases its hundreds digit by 1.
So, 134+100 = **234**.

88. Adding 1 to 555 increases its ones digit by 1.
So, 555+1 = **556**.

89. Adding 10 to 555 increases its tens digit by 1.
So, 555+10 = **565**.

90. Adding 100 to 555 increases its hundreds digit by 1.
So, 555+100 = **655**.

91. Adding 1 to 768 increases its ones digit by 1.
So, 768+1 = **769**.

92. Adding 10 to 768 increases its tens digit by 1.
So, 768+10 = **778**.

93. Adding 100 to 768 increases its hundreds digit by 1.
So, 768+100 = **868**.

94. Since 11 is 1 ten and 1 one, adding 11 to 987 increases
its tens digit by 1 and its ones digit by 1.
So, 987+11 = **998**.

95. Since 101 is 1 hundred and 1 one, adding 101 to 798 increases its hundreds digit by 1 and its ones digit by 1. So, 798+101 = **899**.

96. Since 110 is 1 hundred and 1 ten, adding 110 to 879 increases its hundreds digit by 1 and its tens digit by 1. So, 879+110 = **989**.

97. 11 is 1 ten and 1 one. So, adding 11 to 729 increases its tens digit by 1, and adds 1 more.

Increasing the tens digit of 729 by 1 gives 739, and 1 more is 740. So, 729+11 = **740**.

98. 101 is 1 hundred and 1 one. So, adding 101 to 279 increases its hundreds digit by 1, and adds 1 more.

Increasing the hundreds digit of 279 by 1 gives 379, and 1 more is 380. So, 279+101 = **380**.

99. To add 110 to a number, we add 1 hundred and 1 ten. Adding 1 hundred to 792 gives 892. We cannot add 1 ten to 892 by increasing its tens digit by 1.

But, since 890 is the same as 89 tens, 892 is 89 tens and 2 ones. Adding 1 ten gives 90 tens and 2 ones, which is 902.

So, 792+110 = **902**.

100. Subtracting 1 from 615 decreases its ones digit by 1. So, 615−1 = **614**.

101. Subtracting 10 from 615 decreases its tens digit by 1. So, 615−10 = **605**.

102. Subtracting 100 from 615 decreases its hundreds digit by 1. So, 615−100 = **515**.

103. Subtracting 1 from 888 decreases its ones digit by 1. So, 888−1 = **887**.

104. Subtracting 10 from 888 decreases its tens digit by 1. So, 888−10 = **878**.

105. Subtracting 100 from 888 decreases its hundreds digit by 1. So, 888−100 = **788**.

106. Subtracting 1 from 432 decreases its ones digit by 1. So, 432−1 = **431**.

107. Subtracting 10 from 432 decreases its tens digit by 1. So, 432−10 = **422**.

108. Subtracting 100 from 432 decreases its hundreds digit by 1. So, 432−100 = **332**.

109. 11 is 1 ten and 1 one. So, subtracting 11 from 543 decreases its tens digit by 1 and its ones digit by 1. So, 543−11 = **532**.

110. 101 is 1 hundred and 1 one. So, subtracting 101 from 354 decreases its hundreds digit by 1 and its ones digit by 1. So, 354−101 = **253**.

111. 110 is 1 hundred and 1 ten. So, subtracting 110 from 435 decreases its hundreds digit by 1 and its tens digit by 1. So, 435−110 = **325**.

112. Counting down from 120, we get 120−1 = **119**.

113. Subtracting 100 from 102 decreases its hundreds digit by 1. This leaves 0 hundreds, 0 tens, and 2 ones, which is 2.

So, 102−100 = **2**.

114. We cannot subtract 10 from 201 by decreasing its tens digit by 1. But, since 200 is the same as 20 tens, 201 is 20 tens and 1 one. Taking away 1 ten leaves 19 tens and 1 one, which is 191.

So, 201−10 = **191**.

115. For the first four steps, we increase the tens digit by 1.

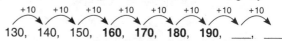

130, 140, 150, **160**, **170**, **180**, **190**, ___, ___.

We cannot increase the tens digit of 190. Since 190 is 19 tens, adding 1 ten to 190 gives 20 tens, which is 200. Then, we increase the tens digit by 1 again to get 210.

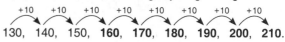

130, 140, 150, **160**, **170**, **180**, **190**, **200**, **210**.

116. For the first two steps, we increase the tens digit by 1.

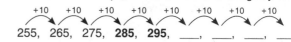

255, 265, 275, **285**, **295**, ___, ___, ___, ___.

Then, since 290 is 29 tens, 295 is 29 tens and 5 ones. Adding 1 ten gives 30 tens and 5 ones, which is 305.

For the last three steps, we increase the tens digit by 1.

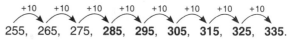

255, 265, 275, **285**, **295**, **305**, **315**, **325**, **335**.

117. Since 580 is 58 tens, 582 is 58 tens and 2 ones.

Adding 1 ten gives 59 tens and 2 ones, which is 592. Adding 1 more ten gives 60 tens and 2 ones, or 602.

From 602, we continue increasing the tens digit to get 612 and 622.

___, ___, 562, 572, 582, **592**, **602**, **612**, **622**.

To the *left* of 562, we *decrease* the number of tens by 1 to fill in 552 and 542 as shown.

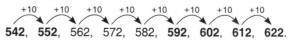

542, **552**, 562, 572, 582, **592**, **602**, **612**, **622**.

118. Since 420 is 42 tens, 424 is 42 tens and 4 ones.

Adding 1 ten gives 43 tens and 4 ones, which is 434. Adding 1 more ten gives 44 tens and 4 ones, or 444.

___, ___, ___, ___, ___, ___, 424, **434**, **444**.

To the *left* of 424, we *decrease* the number of tens by 1.

Subtracting 1 ten gives 41 tens and 4 ones, which is 414. Subtracting 1 more ten gives 40 tens and 4 ones, or 404. Subtracting 1 more ten gives 39 tens and 4 ones, or 394.

We continue the pattern to the left by decreasing the number of tens by 1 to fill in 384, 374, and 364 as shown.

364, **374**, **384**, **394**, **404**, **414**, 424, **434**, **444**.

119. 800 is 8 hundreds, 0 tens, and 0 ones. Breaking a hundred into 10 tens gives 7 hundreds, 10 tens, and 0 ones. Taking away 1 ten leaves 7 hundreds, 9 tens, and 0 ones, which is **790**.

703 is 7 hundreds, 0 tens, and 3 ones.
Breaking a hundred into 10 tens gives 6 hundreds,
10 tens, and 3 ones.
Taking away 1 ten leaves 6 hundreds, 9 tens, and 3 ones,
which is **693**.

309 is 3 hundreds, 0 tens, and 9 ones.
Breaking a hundred into 10 tens gives 2 hundreds,
10 tens, and 9 ones.
Taking away 1 ten leaves 2 hundreds, 9 tens, and 9 ones,
which is **299**.

— *or* —

8 hundreds is 80 tens. Taking away 1 ten leaves 79 tens,
which is **790**.

7 hundreds is the same as 70 tens. So, 703 is 70 tens
and 3 ones. Taking away 1 ten leaves 69 tens and
3 ones, which is **693**.

3 hundreds is the same as 30 tens. So, 309 is 30 tens
and 9 ones. Taking away 1 ten leaves 29 tens and
9 ones, which is **299**.

120. We cannot subtract 10 from 506 by decreasing its tens
digit by 1. But, 5 hundreds is the same as 50 tens. So,
506 is 50 tens and 6 ones. Taking away 1 ten leaves
49 tens and 6 ones, which is 496.

So, there are **496** chipskunks left in the field.

121. We cannot subtract 10 from 405 by decreasing its tens
digit by 1. But, 4 hundreds is the same as 40 tens. So,
405 is 40 tens and 5 ones. Taking away 1 ten leaves
39 tens and 5 ones, which is 395.

So, Captain Kraken has **395** lollipops left in his treasure.

122. We work backwards. Winnie's number is 209. Alex's
number is 100 less than Winnie's number, so Alex's
number is 209 − 100 = 109.

Lizzie's number is 10 less than Alex's number.
So, Lizzie's number is 109 − 10.

1 hundred and 9 ones is the same as 10 tens and 9 ones.
Taking away 1 ten leaves 9 tens and 9 ones, which is 99.
So, Lizzie's number is 99.

Grogg's number is 1 less than Lizzie's number. So,
Grogg's number is 99 − 1 = **98**.

PLACE VALUE
Digit Difference Grids 28–31

123. To get from 63 to 65, we add 1
twice. So, the number in the empty
square is 63 + 1 = **64**.

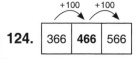

124.

125.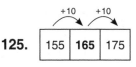

126.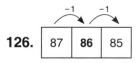

127.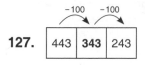

128.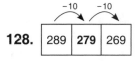

129.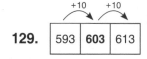

130.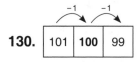

131.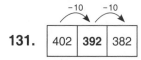

132.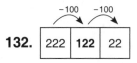

133. To get from 62 to 82, we add 10 twice. So, the number in the
square between 62 and 82 is 62 + 10 = **72**.

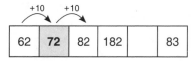

To get from 182 to 83, we subtract 100 and add 1. So,
the number in the square between 182 and 83 is either
182 − 100 = 82 or 182 + 1 = 183.

Since 82 already appears in the grid, we write **183**
between 182 and 83.

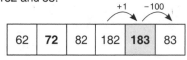

134. To get from 91 to 100, we add 10 and subtract 1. So,
the number in the square between 91 and 100 is either
91 + 10 = 101 or 91 − 1 = 90.

Since 90 already appears in the grid, we write **101** in the
empty square between 91 and 100.

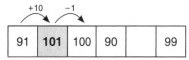

To get from 90 to 99, we add 10 and subtract 1. So,
the number in the square between 90 and 99 is either
90 + 10 = 100 or 90 − 1 = 89.

Since 100 already appears in the grid, we write **89** in the
empty square.

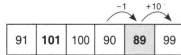

135. To get from 17 to 28, we add 10 and add 1. So, the number in the empty square is either $17+1=18$ or $17+10=27$. Since 27 already appears in the grid, we write **18** in the empty square.

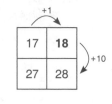

136.

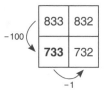

137.

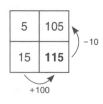

143.

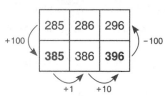

144. Step 1:   Final:

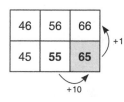

138.

139.

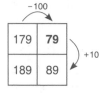

145. Step 1:   Final:

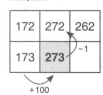

140.

146. Step 1:   Final:

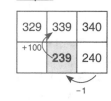

147. Step 1:   Step 2:

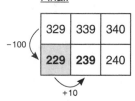

Final:

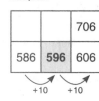

141. To get from 9 to 20, we add 1 and add 10. So, the number in the bottom-left square is either $9+1=10$ or $9+10=19$. Since 10 already appears in the grid, we write **19** in the bottom-left square.

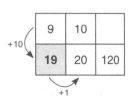

To get from 10 to 120, we add 10 and add 100. So, the number in the top-right square is either $10+10=20$ or $10+100=110$. Since 20 already appears in the grid, we write **110** in the top-right square.

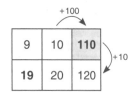

142.

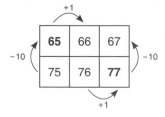

148. Step 1:   Step 2:

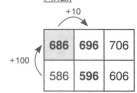

Final:

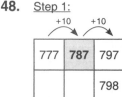

149. Step 1:

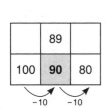

Final:

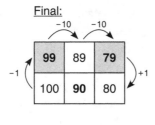

150. To get from 103 to 13, we subtract 100 and add 10. So, the two shaded squares below are 103−100=3 and 103+10=113.

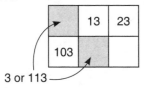

3 or 113

If we write 3 in the bottom-center square, then the bottom-right square must be 13. But, 13 already appears in the grid, so we cannot use this placement.

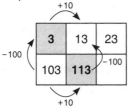

So, we write **3** in the top-left square and **113** in the bottom-center square.

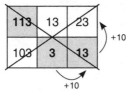

Then, to get from 113 to 23, we add 10 and subtract 100. So, the bottom-right square is 113+10=123 or 113−100=13. Since 13 already appears in the grid, we write **123** in the bottom-right square.

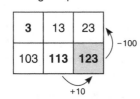

151. We first look at the top-center square that touches 114, 24, and 125.

To get from 114 to 24, we add 10 and subtract 100. So, we could write 124 or 14 in the top-center square. But, 14 is not 1, 10, or 100 apart from 125. Since 124+1=125, we write **124** in the top-center square.

Then, we fill in the remaining squares.

Step 1:   Final:

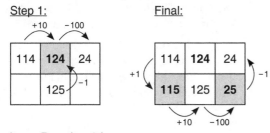

152. From 146 to 66, we subtract 100, add 10, and add 10. There are three different ways we could order these:

−100, +10, +10;
+10, −100, +10;
+10, +10, −100.

We look at a path between 146 and 66, and try all three:

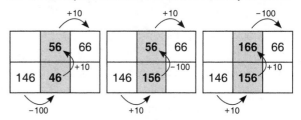

For each of these, we try to fill the remaining squares.

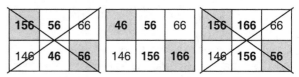

Two of the grids must use the same number twice. So, we cross out those grids out and keep the only solution.

46	56	66
146	156	166

153. Adding 5 hundreds to 213 increases its hundreds digit by 5. So, 213+500=**713**.

154. Adding 5 ones to 213 increases its ones digit by 5. So, 213+5=**218**.

155. Adding 5 tens to 213 increases its tens digit by 5. So, 213+50=**263**.

156. Subtracting 4 ones from 586 decreases its ones digit by 4. So, 586−4=**582**.

157. Subtracting 4 hundreds from 586 decreases its hundreds digit by 4. So, 586−400=**186**.

158. Subtracting 4 tens from 586 decreases its tens digit by 4. So, 586−40=**546**.

159. Adding 7 tens to 121 increases its tens digit by 7. So, 121+70=**191**.

160. Adding 7 ones to 121 increases its ones digit by 7. So, 121+7=**128**.

161. Adding 7 hundreds to 121 increases its hundreds digit by 7. So, 121+700=**821**.

162. Subtracting 3 hundreds from 475 decreases its hundreds digit by 3. So, 475−300=**175**.

163. Subtracting 3 tens from 475 decreases its tens digit by 3. So, 475−30=**445**.

164. Subtracting 3 ones from 475 decreases its ones digit by 3. So, 475−3=**472**.

165. 4 tens and 4 ones plus 1 ten and 7 ones is $4+1=5$ tens and $4+7=11$ ones. We cannot write an 11 in the ones place. So, we regroup 10 ones to make 1 ten.

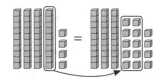

This gives 6 tens and 1 one, which is 61. So, $44+17=\textbf{61}$.

166. We cannot subtract 7 ones from 4 ones. So, we break 1 ten into 10 ones.

So, 44 is the same as 3 tens and 14 ones. Subtracting 1 ten and 7 ones leaves $3-1=2$ tens and $14-7=7$ ones.

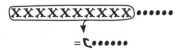

So, $44-17=\textbf{27}$.

167. There are ten $\mathbf{X}$'s and six •'s in all.

$$\mathbf{XXXXXXXXXX}••••••$$

We replace ten $\mathbf{X}$'s with one $\mathbf{C}$.

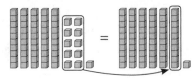

So, we have

$$\mathbf{XXXX}•• + \mathbf{XXXXXX}•••• = \mathbf{C}•••••• = 106.$$

So, $42+64=\textbf{106}$.

168. We cannot take away four $\mathbf{X}$'s from one $\mathbf{X}$. So, we break the $\mathbf{C}$ in $\mathbf{CX}••$ into ten $\mathbf{X}$'s.

$$\mathbf{CX}•• = \mathbf{XXXXXXXXXXX}••$$

Taking away four $\mathbf{X}$'s leaves $\mathbf{XXXXXXX}•• = 72$.

So, $112-40=\textbf{72}$.

169. 683 is 6 hundreds, 8 tens, and 3 ones. Adding 7 tens to 683 gives 6 hundreds, $8+7=15$ tens, and 3 ones.

We regroup 10 tens into 1 hundred. This gives 7 hundreds, 5 tens, and 3 ones, or 753.

So, $683+70=\textbf{753}$.

— *or* —

680 is 68 tens. So, 683 is 68 tens and 3 ones. Adding 7 tens gives $68+7=75$ tens and 3 ones, which is **753**.

170. 446 is 4 hundreds, 4 tens, and 6 ones. We cannot take away 8 tens from 4 tens.

We break 1 hundred into 10 tens. This gives 3 hundreds, 14 tens, and 6 ones.

Then, subtracting 8 tens leaves 3 hundreds, $14-8=6$ tens, and 6 ones, which is 366.

So, $446-80=\textbf{366}$.

— *or* —

440 is 44 tens. So, 446 is 44 tens and 6 ones. Taking away 8 tens leaves $44-8=36$ tens and 6 ones, which is **366**.

171. 111 is 1 hundred, 1 ten, and 1 one. So, adding 111 to 789 gives 8 hundreds, 9 tens, and 10 ones.

We regroup 10 ones to make 1 ten. This gives 8 hundreds, 10 tens, and 0 ones.

Then, we regroup 10 tens to make 1 hundred. This gives 9 hundreds, 0 tens, and 0 ones, which is 900.

So, 111 more than 789 is **900**.

172. 0 cannot be the leftmost digit of a three-digit number. So, the leftmost digit must be 1 or 2.

If 1 is the leftmost digit, the other two digits are 0 and 2. There are two ways to order them: 102 and 120.

If 2 is the leftmost digit, the other two digits are 0 and 1. There are two ways to order them: 201 and 210.

So, the four different three-digit numbers that use the digits 0, 1, and 2 are **102**, **120**, **201**, and **210**.

173. 789 is 7 hundreds, 8 tens, and 9 ones.

7 hundreds means there are 7 $\mathbf{C}$'s in the pirate number. 8 tens means there are 8 $\mathbf{X}$'s in the pirate number. 9 ones means there are 9 •'s in the pirate number.

So, there are $7+8+9=\textbf{24}$ symbols in the pirate number for 789.

174. Since we are writing a 3-digit number, the number must be at least 100, and use at least one $\mathbf{C}$. We can use any of the three pirate symbols as the second symbol. So, we have $\mathbf{C}•$, $\mathbf{CX}$, and $\mathbf{CC}$.

Using digits, these numbers are 101, 110, and 200.

175. 358 is 35 tens and 8 ones. To fill the missing blanks, we either count up or down by 3 tens from 35 tens. Adding 3 tens to a number does not change its ones digit.

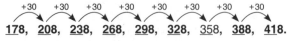

176. 77 tens plus 77 ones is the same as 77 tens plus 7 tens and 7 ones. This gives $77+7=84$ tens and 7 ones. 84 tens is 840, so 84 tens and 7 ones is **847**.

177. To make the smallest three-digit number possible, we start by making the hundreds digit as small as possible. The hundreds digit of a three-digit number cannot be 0. So, we consider numbers with hundreds digit 1.

$$1 \_\_ \_\_$$

Then, we make the tens digit as small as possible. Since the sum of all three digits is 12, the tens and ones digits must have a sum of $12-1=11$.

If the tens digit is 0, then the ones digit must be 11. But, we cannot write 11 in the ones place.

If the tens digit is 1, then the ones digit must be 10. But, we cannot write 10 in the ones place.

If the tens digit is 2, then the ones digit must be 9. This works!

So, **129** is the smallest three-digit number with a digit sum of 12.

178. When writing numbers between 1 and 100, the digit 5 can only appear in the ones or the tens place.

The numbers less than 100 with ones digit 5 are:

5, 15, 25, 35, 45, 55, 65, 75, 85, 95.

So, the digit 5 appears ten times in the ones place.

The numbers less than 100 with tens digit 5 are:

50, 51, 52, 53, 54, 55, 56, 57, 58, 59.

So, the digit 5 appears ten times in the tens place.

So, when writing every number from 1 to 100, you will write the digit 5 a total of $10+10=$ **20** times.

Note that there are only 19 numbers that include the digit 5, but since 55 has two fives, we write the digit 5 20 times.

COMPARING

The Number Line 37-39

1.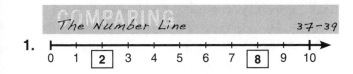

2.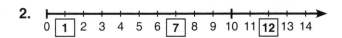

3.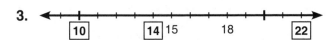

4.

5.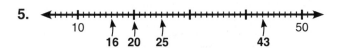

6.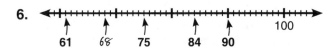

7.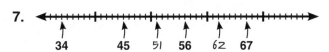

8. We move two tick marks to get from 80 to 100.
So, the tick marks on this number line count by 10's.

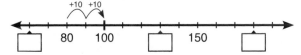

We count by 10's to fill in the boxes as shown below.

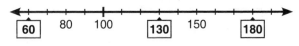

9. We move six tick marks to get from 320 to 380.
So, the tick marks on this number line count by 10's.

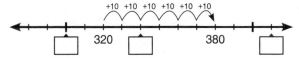

We count by 10's to fill in the boxes as shown below.

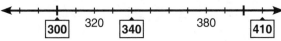

10. We move three tick marks to get from 150 to 300.
So, the tick marks on this number line count by 50's.

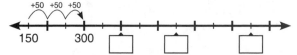

We count by 50's to fill in the boxes as shown below.

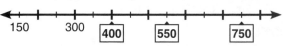

11. Before filling the boxes, it is helpful to label the big tick marks. The tick marks on this number line count by 50's, so the big tick marks are 50+50 = 100 units apart.

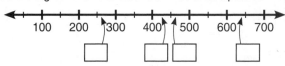

There are six 3-digit numbers we can make using the digits 2, 4, and 6 once each:

246, 264, 426, 462, 624, and 642.

The only number in our list between 250 and 300 is 264.
The only number in our list between 400 and 450 is 426.
The only number in our list between 450 and 500 is 462.
Both 624 and 642 are between 600 and 650.
Since the arrow points much closer to 650 than to 600, we label the rightmost arrow with 642.

We label these four numbers as shown below.

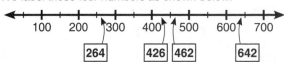

COMPARING

Distance 40-41

12. From 19 to 20 is 1 unit.
From 20 to 60 is 10+10+10+10 = 40 units.

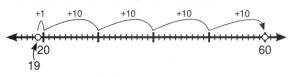

So, the distance from 19 to 60 is 1+40 = **41** units.

13. The left dot is on the number 84, and the right dot is on the number 124.

From 84 to 90 is 6 units.
From 90 to 120 is 10+10+10 = 30 units.
From 120 to 124 is 4 units.

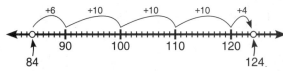

So, the distance between the two dots is 6+30+4 = **40** units.

— *or* —

84 and 124 have the same ones digit. So, we can count up by 10's to get from 84 to 124.

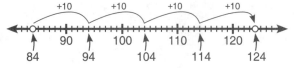

So, the distance between the two dots is
$10+10+10+10 = **40**$ units.

14. From 54 to 60 is 6 units.
From 60 to 80 is $10+10 = 20$ units.
From 80 to 81 is 1 unit.

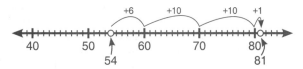

So, the distance from 54 to 81 is $6+20+1 = **27**$ units.

— *or* —

We count up by 10's to get from 54 to 74. Then, we count up by 7 more to get to 81.

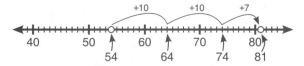

So, the distance from 54 to 81 is $10+10+7 = **27**$ units.

15. The distance from 18 to 25 is $2+5 = 7$ units.

To find the other number that is 7 units from 25, we count up 7 units from 25.

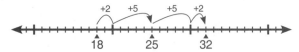

So, 18 and **32** are the same distance from 25.

16. The distance from 33 to 50 is $7+10 = 17$ units.

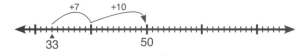

To find the other number that is 17 units from 50, we count up 17 units from 50.

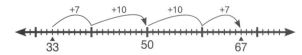

So, 33 and **67** are the same distance from 50.

17. The distance from 73 to 98 is $10+10+5 = 25$ units.

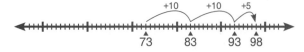

To find the other number that is 25 units from 73, we count down 25 units from 73.

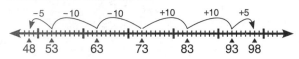

So, 98 and **48** are the same distance from 73.

18. We count in from 45 and 65.

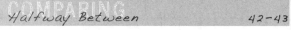

The number halfway between 45 and 65 is the same as the number halfway between 50 and 60.

The number halfway between 50 and 60 is 55.
So, the number halfway between 45 and 65 is **55**.

19. We count in from 210 and 260.

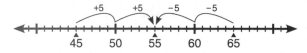

The number halfway between 210 and 260 is the same as the number halfway between 230 and 240.

The number halfway between 230 and 240 is 235.
So, the number halfway between 210 and 260 is **235**.

20. We count in from 27 and 73.

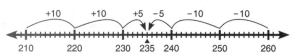

The number halfway between 27 and 73 is the same as the number halfway between 30 and 70.

The number halfway between 30 and 70 is 50.
So, the number halfway between 27 and 73 is **50**.

21. If we try to count in to 40 and 70, then we must count in by different amounts.

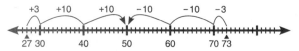

When finding the number halfway between two other numbers, we must always count in by the same amount. So, we try something else.

Since it is easy to add and subtract 10's, we try counting in by 10's from 32 and 76.

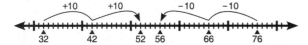

The number halfway between 32 and 76 is the same as the number halfway between 52 and 56.

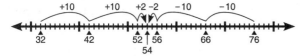

The number halfway between 52 and 56 is 54.
So, the number halfway between 32 and 76 is **54**.

22. We count in by 10's from 534 and 584.

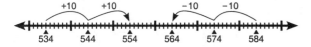

The number halfway between 534 and 584 is the same as the number halfway between 554 and 564.

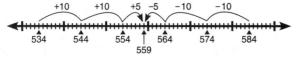

The number halfway between 554 and 564 is 559.
So, the number halfway between 534 and 584 is **559**.

23. We count in by 10's from 70 and 130.

The number halfway between 70 and 130 is the same as the number halfway between 80 and 120, which is the number halfway between 90 and 110, which is 100.

So, the number halfway between 70 and 130 is **100**.

24. We count in by 2 from 118 and 162.

The number halfway between 118 and 162 is the same as the number halfway between 120 and 160.

Then, we count in by 10's from 120 and 160.

The number halfway between 120 and 160 is the same as the number halfway between 130 and 150, which is 140.

So, the number halfway between 118 and 162 is **140**.

25. We count in by 10's from 43 and 89.

The number halfway between 43 and 89 is the same as the number halfway between 53 and 79, which is the number halfway between 63 and 69.

The number halfway between 63 and 69 is 66.

So, the number halfway between 43 and 89 is **66**.

26. To find out if 64 is closer to 40 or to 90, we find the number that is halfway between 40 and 90.

The number halfway between 40 and 90 is the same as the number halfway between 50 and 80, which is the number halfway between 60 and 70, which is 65.

64 is less than 65. So, 64 is closer to **40** than it is to 90.

— *or* —

The distance from 40 to 64 is $10+10+4 = 24$ units.
The distance from 64 to 90 is $6+10+10 = 26$ units.

So, 64 is closer to **40** than it is to 90.

27. The number halfway between 350 and 410 is the same as the number halfway between 360 and 400, which is the number halfway between 370 and 390, which is 380.

385 is more than 380. So, 385 is closer to **410** than it is to 350.

— *or* —

The distance from 350 to 385 is $10+10+10+5 = 35$ units.
The distance from 385 to 410 is $5+10+10 = 25$ units.

So, 385 is closer to **410** than it is to 350.

Number Line Conquest   44-45

28. Only by labeling the X shown below can Player X win.

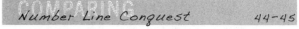

We score the game as shown below.

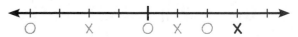

X scores 1 point
O scores 0 points

29. Only by labeling the X shown below can Player X win.

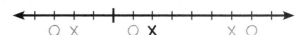

We score the game as shown below.

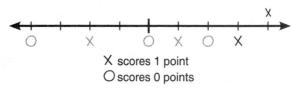

X scores 4 points
O scores 3 points

30. Only by labeling the X shown below can Player X win.

We score the game as shown below.

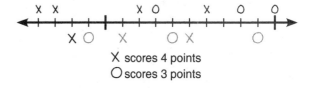

X scores 4 points
O scores 3 points

31. Only by labeling the X shown below can Player X win.

We score the game as shown below.

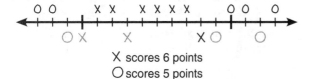

X scores 6 points
O scores 5 points

32. Only by labeling the X shown below can Player X win.

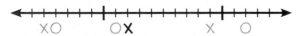

We score the game as shown below.

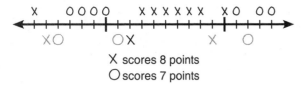

X scores 8 points
O scores 7 points

COMPARING
Honeycomb Paths 46–49

33. We start by placing 2 to connect 1 to 3.

Then, there is only one way to place 5 and 6 to connect 4 to 7.

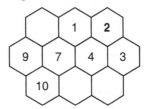

We complete the puzzle by placing 8 to connect 7 to 9.

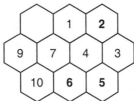

34. There is only one way to place 3 and 4 to connect 2 to 5.

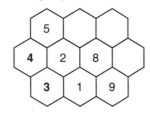

Then, there is only one way to place 6 and 7 to connect 5 to 8.

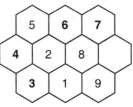

Finally, we complete the puzzle by placing 10 as shown.

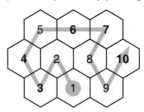

We use the strategies from the previous problems to complete the following puzzles.

35. **36.**

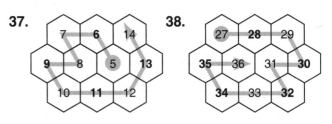

37. **38.**

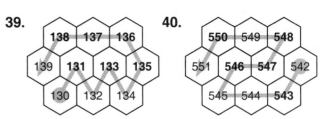

39. **40.**

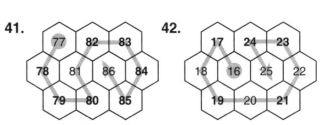

41. **42.**

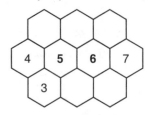

43. There is only one way to place 5 and 6 to connect 4 to 7.

There are two empty hexagons touching 7. One of these hexagons must contain an 8. If we place the 8 below the 7, then our path of consecutive numbers cannot cross every hexagon.

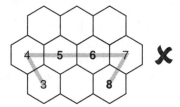

So, we place the 8 above the 7. Then, we fill the empty hexagons to the left of 8 with 9 and 10.

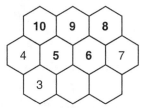

Finally, we can only fill the empty hexagons to the right of 3 with 2 and 1, as shown below.

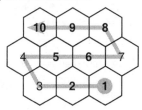

44. 398 and 401 must be connected by 2 empty hexagons. 401 and 405 must be connected by 3 empty hexagons.

There is only one way to connect 398 and 401 with two empty hexagons **and** connect 401 and 405 with three empty hexagons, as shown below.

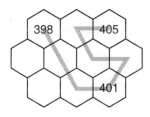

We fill these hexagons as shown.

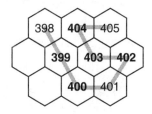

We complete the puzzle by filling the remaining empty hexagons with 397 and 396.

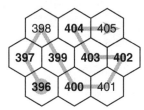

We use the strategies from the previous problems to complete the following puzzles.

45.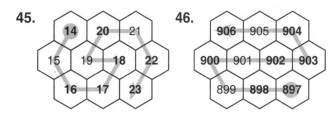

46.

47.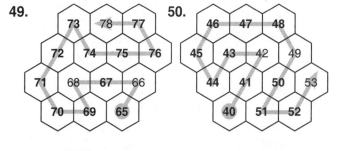

48.

49.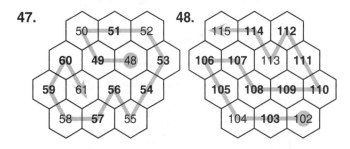

50.

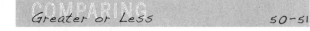

COMPARING

Greater or Less 50-51

51. 98 is greater than 89.
So, we fill the circle with the "is greater than" symbol.

$$98 \gtrdot 89$$

52. 2+7 = 9, and 9 is less than 27.
So, we fill the circle with the "is less than" symbol.

$$2+7 \lessdot 27$$

53. 9+3 and 3+9 both equal 12.
So, we fill the circle with the "is equal to" symbol.

$$9+3 \doteq 3+9$$

54. 40+4 = 44, and 50−5 = 45.
44 is less than 45. So, we fill the circle as shown.

$$40+4 \lessdot 50-5$$

55. 213 is greater than 132. So, we fill the circle as shown.

$$213 \gtrdot 132$$

56. 7+80 = 87, and 70+8 = 78.
87 is greater than 78. So, we fill the circle as shown.

$$7+80 \gtrdot 70+8$$

57. 19 ones is 19, and 2 tens is 20.
19 is less than 20. So, we fill the circle as shown.

$$19 \text{ ones} \lessdot 2 \text{ tens}$$

58. 30 tens is the same as 3 hundreds, or 300. So, we fill the circle as shown.

30 tens $=$ 3 hundreds

59. 199 is 1 less than 200, and 202 is 2 more than 200. So, the sum 199+202 is 1 more than 200+200.

We fill the circle as shown.

199+202 $>$ 200+200

60. 99 is 1 less than 100. So, the sum of three 99's is 3 less than the sum of three 100's.

So, 99+99+99 and 300−3 are equal.

99+99+99 $=$ 300−3

61. 60 is the only two-digit number with tens digit 6 that is less than 61. So, we fill the blank with 0.

6$\boxed{0}$<61

62. Filling the blank with a digit that is 8 or less gives a result that is 84 or less, which is not greater than 85.

So, we can only fill the blank with 9.

$\boxed{9}$4>85

63. We start by looking at 7▢>73. Among the digits 2, 3, and 4, only 4 can fill this blank.

Then, we look at 30<▢1. Among the remaining digits, 2 and 3, only 3 can fill this blank.

Finally, we fill the blank in ▢3<34 with the only remaining digit, 2.

30<$\boxed{3}$1

7$\boxed{4}$>73

$\boxed{2}$3<34

64. We start by looking at 46>4▢. Among the digits 5, 6, 7, and 8, only 5 can fill this blank: 46>4$\boxed{5}$.

Then, we look at 7▢<▢5. The number 7▢ must be 7$\boxed{6}$, 7$\boxed{7}$, or 7$\boxed{8}$. So, the number ▢5 cannot be 75 or less. So, only 8 can fill the second blank: 7▢<$\boxed{8}$5.

Next, we look at ▢1>62. Among the remaining digits 6 and 7, only 7 can fill this blank: $\boxed{7}$1>62.

We fill the remaining blank in 7▢<$\boxed{8}$5 with the only digit we have not yet used, 6. This gives 7$\boxed{6}$<$\boxed{8}$5.

$\boxed{7}$1>62

7$\boxed{6}$<$\boxed{8}$5

46>4$\boxed{5}$

COMPARING
Ordering   52–53

65. 32 is the only number less than 100, so it is the smallest.

32, ___, ___, ___

Next, we have two numbers in the 200's and one number in the 300's. Any number in the 300's is larger than any

number in the 200's. So, 323 is largest.

32, ___, ___, 323

Finally, since 23 is less than 33, we know 223 is less than 233. From least to greatest, we have

32, 223, 233, 323.

66. We have two numbers in the 400's and two numbers in the 700's. Any number in the 400's is smaller than any number in the 700's. So, 471 and 417 are the two smallest numbers.

Since 17 is less than 71, we know 417 is less than 471. So, the first number is 417, followed by 471.

417, 471, ___, ___

Since 14 is less than 41, we know that 714 is less than 741. So, the next number is 714, followed by 741.

From least to greatest, we have

417, 471, 714, 741.

67. To make the largest three-digit number whose digits are all different, we use the three largest digits: 7, 8, and 9.

Any number in the 900's is greater than any number in the 700's or 800's. So, we write 9 in the hundreds place.

9 _ _

We must fill the remaining place values with 7 and 8. Since 987 is greater than 978, the greatest number we can make is **987**.

68. To make the smallest three-digit number whose digits are all different, we use the three smallest digits: 0, 1, and 2.

The hundreds digit cannot be 0, otherwise we get a *two-digit* number, such as 012 (which is just 12).

So, the hundreds digit must be 1 or 2. Any number in the 100's is smaller than any number in the 200's. So, we write 1 in the hundreds place.

1 _ _

We must fill the remaining place values with 0 and 2. Since 102 is less than 120, the smallest number we can make is **102**.

69. Three-digit numbers are greater than two-digit numbers, which are greater than one-digit numbers.

So, we order the three-digit numbers first, then the two-digit numbers, then the one-digit number.

Hundreds	Tens	Ones		Hundreds	Tens	Ones
7	8	5		7	8	5
1	1	5		7	5	8
	5	1		5	1	1
	2	3		2	0	3
		6		1	1	5
7	5	8			5	1
5	1	1			2	3
2	0	3				6

70. We order the three-digit numbers first, then the two-digit numbers, then the one-digit numbers.

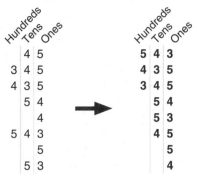

71. We order the three-digit numbers first, then the two-digit numbers, then the one-digit number.

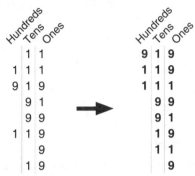

72.
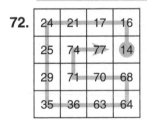

73.

74.
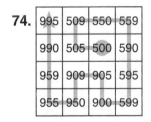

75.
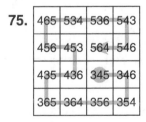

76.
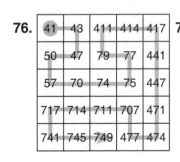

77.
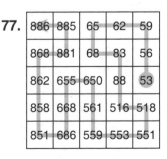

78.

79.

80. We have **16**<**18**<**27**<**61**<**72**<**108**.

81. We have **650**>**605**>**560**>**506**>**65**>**56**.

82. The number that replaces the gray box must be greater than 10 **and** less than 20. Only the whole numbers from 11 to 19 are greater than 10 and less than 20:

11, 12, 13, 14, 15, 16, 17, 18, 19.

So, **9** different numbers could replace the gray box.

83. We have **2**<**9**<**22**<**29**<**92**<**99**<**222**<**229**.

84. We have 33<**34**<43<44<**333**<334<**343**<**344**<433.

85. We have 123<**132**<**213**<**231**<312<**321**.

86. We have 0<5<50<**55**<**500**<**505**<**550**<555.

87. We have 7<8<77<**78**<**87**<**88**<**777**<778.

88. Since the ones digits are all the same, we place 4, 7, and 8 in the tens digit blanks in order from least to greatest.

$$\boxed{4}2<\boxed{7}2<\boxed{8}2$$

89. Since the largest number is 80, we cannot place 9 in either tens digit blank. So, we must place 9 in the leftmost blank.

$$\boxed{9}<\square0<\square0<80$$

Then, we place the remaining digits 3 and 5 as shown.

$$\boxed{9}<\boxed{3}0<\boxed{5}0<80$$

90. Among the given digits, only 6 can be placed in the leftmost blank.

$$34<3\boxed{6}<4\square<4\square$$

Then, we place the remaining digits 2 and 4 as shown.

$$34<3\boxed{6}<4\boxed{2}<4\boxed{4}$$

91. Since the largest number is 60, we cannot place 6 or 7 in either tens digit blank. So, the tens digit blanks must be filled with 4 and 5.

$$\boxed{4}\square<4\square<\boxed{5}4<60$$

Then, we place the digits 6 and 7 as shown.

$$\boxed{4}\boxed{6}<4\boxed{7}<\boxed{5}4<60$$

92. Since the middle number has tens digit 9, the rightmost number must also have tens digit 9.

$$\boxed{}9<9\boxed{}<\boxed{9}7$$

Among the remaining digits 6 and 8, only 6 can be placed in the ones blank of the middle number.

$$\boxed{}9<9\boxed{6}<\boxed{9}7$$

So, we place 8 in the remaining blank.

$$\boxed{8}9<9\boxed{6}<\boxed{9}7$$

93. Since the largest number is 75, we cannot place 7 in either tens digit blank. So, we place the 7 as shown.

$$5<\boxed{7}<\boxed{}5<\boxed{}5<75$$

Then, we place the remaining digits 5 and 6 as shown.

$$5<\boxed{7}<\boxed{5}5<\boxed{6}5<75$$

94. 11 is the smallest number we can make using the given digits, followed by 12, then 21, then 22. So, we fill the blanks as shown.

$$\boxed{1}\boxed{1}<\boxed{1}\boxed{2}<\boxed{2}\boxed{1}<\boxed{2}\boxed{2}$$

95. Since the smallest number is 40 and the largest number is 70, the digits 3, 9, and 9 cannot be placed in any of the tens digit blanks. So, we must place 4, 5, and 5 in the tens digit blanks as shown.

$$40<\boxed{4}\boxed{}<\boxed{5}\boxed{}<\boxed{5}\boxed{}<70$$

Since 59 is not greater than 59, we cannot place both 9's in the blanks with tens digit 5. So, we place the 3, 9, and 9 as shown.

$$40<\boxed{4}\boxed{9}<\boxed{5}\boxed{3}<\boxed{5}\boxed{9}<70$$

96. We try working from the outside in. The smallest number we can make using the given digits is 55, and the largest number we can make is 99. So, we try filling the blanks as shown below.

Digits: 5̶, 5̶, 5, 5, 7, 7, 9, 9̶, 9̶

$$\boxed{5}\boxed{5}<\boxed{}7<\boxed{}\boxed{}<7\boxed{}<\boxed{}7<\boxed{9}\boxed{9}$$

The next-smallest number we can make is 57, and the next-largest number we can make is 97. So, we try the following.

Digits: 5̶, 5̶, 5̶, 5, 7, 7, 9̶, 9̶, 9̶

$$\boxed{5}\boxed{5}<\boxed{5}7<\boxed{}\boxed{}<7\boxed{}<\boxed{9}7<\boxed{9}\boxed{9}$$

We fill the remaining blanks as shown below.

Digits: 5̶, 5̶, 5̶, 5̶, 7̶, 7̶, 9̶, 9̶, 9̶

$$\boxed{5}\boxed{5}<\boxed{5}7<\boxed{7}\boxed{5}<7\boxed{7}<\boxed{9}7<\boxed{9}\boxed{9}$$

This is the only solution.

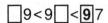

97. There are many ways to count in from 123 and 321. For example, we could count in by 21, then by 56, then by 20 as shown below.

The number halfway between 123 and 321 is the same as the number halfway between 220 and 224, which is 222.

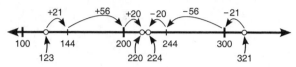

So, the number halfway between 123 and 321 is **222**.

— *or* —

We count in by 100 from 123 and 321.

The number halfway between 123 and 321 is the same as the number halfway between 223 and 221.

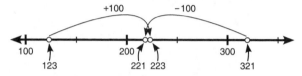

The number halfway between 223 and 221 is 222.

So, the number halfway between 123 and 321 is **222**.

98. We move three tick marks to get from 15 to 60. So, the tick marks on this number line count by 15's.

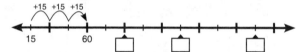

We move two tick marks to get from one big tick mark to the next. So, the big tick marks are $15+15=30$ units apart.

We count by 30's to label the big tick marks as shown.

Then, we fill the remaining boxes with $120+15=135$ and $180+15=195$.

99. We label Alex's and Lizzie's numbers on the number line.

The distance between Alex's number and Lizzie's number is $10+10+10=30$ units.

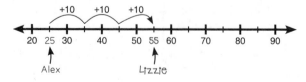

Lizzie's number is halfway between Alex's number and Grogg's number. So, the distance between Lizzie's number and Grogg's number is also 30 units.

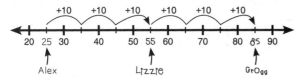

So, Grogg's number is **85**.

100. Ms. Q marks the number 50, and Alex marks the number 15 units to the left of Ms. Q's number. So, Alex marks 35.

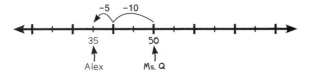

Grogg marks the number 25 units to the right of Alex's number. So, Grogg marks 60.

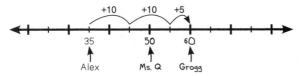

Lizzie marks the number 35 units to the left of Grogg's number. So, Lizzie marks 25.

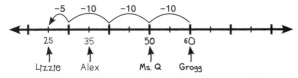

Winnie marks the number 45 units to the right of Lizzie's number. So, Winnie marks **70**.

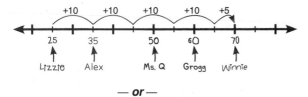

— *or* —

Moving left 15 units then right 25 units is the same as moving right 10 units. So, Grogg's number is 10 units to the right of Ms. Q's number.

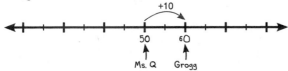

Moving left 35 units then right 45 units is the same as moving right 10 units. So, Winnie's number is 10 units to the right of Grogg's number.

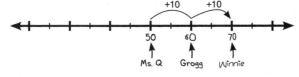

So, Winnie marks **70**.

101. The largest number has hundreds digit 3. Since the △ is the hundreds digit of the smallest number, we know the △ is 3 or less. But, since △■●<3■●, the △ cannot be 3. So, the △ is 2 or less.

Looking at the tens digits in 3●■<3△△, we know the ● is less than the △.

Looking at the tens digits in 3■●<3●■, we know the ■ is less than the ●.

So, we have ■<●<△, and the △ is 2 or less. This leaves only one possible value for each shape.

We have △ = **2**, ● = **1**, and ■ = **0**.

$$△■●<3■●<3●■<3△△$$
$$201<301<310<322$$

102. We start by looking at □7□<□□0. Since each blank is filled with the same digit, both □7□ and □□0 have the same hundreds digit.

So, for □□0 to be greater than □7□, the tens digit of □□0 must be more than 7. So, the digit in each blank is either 8 or 9.

Next, we look at □□0<9□0. If the digit in each blank is 9, then □□0 will be equal to 9□0.

So, the digit Alba uses in each blank is **8**.

$$\boxed{8}7\boxed{8}<\boxed{8}\boxed{8}0<9\boxed{8}0$$

103. Since Tia's favorite number is between 777 and 999, the hundreds digit of her number is 7, 8, or 9.

The hundreds digit is less than the ones digit, which is less than the tens digit. This means that the hundreds digit is at least *two* less than the tens digit.

So, the hundreds digit cannot be 8 or 9 and must be 7.

Since the hundreds digit of Tia's number is 7, the ones digit is 8, and the tens digit is 9.

So, Tia's favorite number is **798**.

104. The smallest number that could be under ⬣ is 16.
So, the smallest number that could be under ▲ is 17.

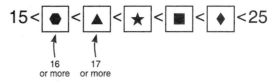

The largest number that could be under ◆ is 24.
So, the largest number that could be under ■ is 23.

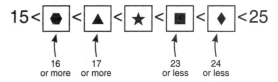

So, the number under ★ must be more than 17 *and* less than 23. Only the whole numbers from 18 to 22 are more than 17 and less than 23:

18, 19, 20, 21, 22.

105. There are 9 different two-digit numbers we can make using the digits 1, 3, and 6. We label these numbers on the number line below.

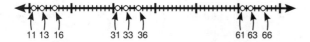

11 13 16   31 33 36   61 63 66

We can rule out any number in the 10's or 60's as the middle number, so the middle number is in the 30's. This only works if the smallest number is in the 10's, and the largest number is in the 60's.

So, we have

⬚3⬚ is halfway between ⬚1⬚ and ⬚6⬚.

Using 1's, 3's, and 6's, there is only one way to fill the ones digits of these numbers to make a true statement:

⬚3⬚6 is halfway between ⬚1⬚1 and ⬚6⬚1.

<u>Check</u>:
The distance between 11 and 36 is 25 units.
The distance between 36 and 61 is 25 units. ✓

106.

7⃞7 < 6⃞6 < 5⃞5 ✗   3⃞3 < 2⃞9 < 3⃞2 ✗
5⃞7 < 6⃞6 < 7⃞5 ✓   2⃞3 < 2⃞9 < 3⃞3 ✓

1⃞2 < 1⃞2 < 1⃞2 ✗   2⃞2 < 5⃞5 < 2⃞2 ✗
1⃞1 < 1⃞2 < 2⃞2 ✓   2⃞2 < 2⃞5 < 5⃞2 ✓

ADDITION

Sums 65-67

1. Adding 32+56 gives 3+5 = 8 tens and 2+6 = 8 ones. 8 tens is 80, so 8 tens and 8 ones is 80+8 = 88.

$$32+56 = \underline{80} + \underline{8} = \underline{\textbf{88}}.$$

2. Adding 13+51 gives 1+5 = 6 tens and 3+1 = 4 ones. 6 tens is 60, so 6 tens and 4 ones is 60+4 = 64.

$$13+51 = \underline{\textbf{60}} + \underline{4} = \underline{\textbf{64}}.$$

3. Adding 62+34 gives 6+3 = 9 tens and 2+4 = 6 ones. 9 tens is 90, so 9 tens and 6 ones is 90+6 = 96.

$$62+34 = \underline{90} + \underline{\textbf{6}} = \underline{\textbf{96}}.$$

4. Adding 54+21 gives 5+2 = 7 tens and 4+1 = 5 ones. 7 tens is 70, so 7 tens and 5 ones is 70+5 = 75.

$$54+21 = \underline{\textbf{70}} + \underline{\textbf{5}} = \underline{\textbf{75}}.$$

5. Adding 217+461 gives 2+4 = 6 hundreds, 1+6 = 7 tens, and 7+1 = 8 ones. 6 hundreds, 7 tens, and 8 ones is 600+70+8 = 678.

$$217+461 = \underline{600} + \underline{\textbf{70}} + \underline{8} = \underline{\textbf{678}}.$$

6. Adding 47+32 gives 4+3 = 7 tens and 7+2 = 9 ones. 7 tens and 9 ones is **79**.

7. Adding 61+25 gives 6+2 = 8 tens and 1+5 = 6 ones. 8 tens and 6 ones is **86**.

8. Adding 136+53 gives 1 hundred, 3+5 = 8 tens, and 6+3 = 9 ones. 1 hundred, 8 tens, and 9 ones is **189**.

9. Adding 64+315 gives 3 hundreds, 6+1 = 7 tens, and 4+5 = 9 ones. 3 hundreds, 7 tens, and 9 ones is **379**.

10. Adding 250+37 gives 2 hundreds, 5+3 = 8 tens, and 0+7 = 7 ones. 2 hundreds, 8 tens, and 7 ones is **287**.

11. Adding 270+604 gives 2+6 = 8 hundreds, 7+0 = 7 tens, and 0+4 = 4 ones. 8 hundreds, 7 tens, and 4 ones is **874**.

12. Adding 316+260 gives 3+2 = 5 hundreds, 1+6 = 7 tens, and 6+0 = 6 ones. 5 hundreds, 7 tens, and 6 ones is **576**.

13. $43+83 = \underline{120} + \underline{6} = \underline{\textbf{126}}.$

14. $72+46 = \underline{110} + \underline{8} = \underline{\textbf{118}}.$

15. $54+19 = \underline{60} + \underline{\textbf{13}} = \underline{\textbf{73}}.$

16. $36+28 = \underline{50} + \underline{\textbf{14}} = \underline{\textbf{64}}.$

17. $71+83 = \underline{\textbf{150}} + \underline{4} = \underline{\textbf{154}}.$

18. $59+39 = \underline{80} + \underline{\textbf{18}} = \underline{\textbf{98}}.$

19. $52+81 = \underline{\textbf{130}} + \underline{3} = \underline{\textbf{133}}.$

20. $67+28 = \underline{80} + \underline{\textbf{15}} = \underline{\textbf{95}}.$

21. $38+83 = \underline{\textbf{110}} + \underline{\textbf{11}} = \underline{\textbf{121}}.$

22. $85+65 = \underline{\textbf{140}} + \underline{\textbf{10}} = \underline{\textbf{150}}.$

23. $223+195 = \underline{300} + \underline{110} + \underline{8} = \underline{\textbf{418}}.$

24. $528+345 = \underline{800} + \underline{60} + \underline{\textbf{13}} = \underline{\textbf{873}}.$

25. $393+461 = \underline{700} + \underline{150} + \underline{4} = \underline{\textbf{854}}.$

26. $315+167 = \underline{\textbf{400}} + \underline{\textbf{70}} + \underline{\textbf{12}} = \underline{\textbf{482}}.$

27. $248+590 = \underline{\textbf{700}} + \underline{\textbf{130}} + \underline{8} = \underline{\textbf{838}}.$

28. $273+64 = \underline{\textbf{200}} + \underline{\textbf{130}} + \underline{7} = \underline{\textbf{337}}.$

29. $883+109 = \underline{\textbf{900}} + \underline{\textbf{80}} + \underline{\textbf{12}} = \underline{\textbf{992}}.$

30. $555+346 = \underline{\textbf{800}} + \underline{\textbf{90}} + \underline{\textbf{11}} = \underline{\textbf{901}}.$

ADDITION
Cross-Number Puzzles 68-69

31.

10	+	11	=	21
+		+		+
15	+	12	=	27
=		=		=
25	+	23	=	48

32.

70	+	20	=	90
+		+		+
110	+	40	=	150
=		=		=
180	+	60	=	240

33.

40	+	36	=	76
+		+		+
37	+	20	=	57
=		=		=
77	+	56	=	133

34.

22	+	34	=	56
+		+		+
45	+	41	=	86
=		=		=
67	+	75	=	142

35.

127	+	60	=	187
+		+		+
40	+	172	=	212
=		=		=
167	+	232	=	399

36.

32	+	30	=	62
+		+		+
41	+	44	=	85
=		=		=
73	+	74	=	147

37.

120	+	70	=	190
+		+		+
30	+	40	=	70
=		=		=
150	+	110	=	260

38.

191	+	301	=	492
+		+		+
123	+	212	=	335
=		=		=
314	+	513	=	827

ADDITION
Number Line 70-71

39. 196+36 = 232.

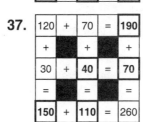

40. $880+88=968$.

41. $350+57=407$.

42. $175+29=204$.

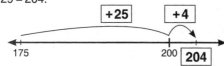

43. $391+39=430$.

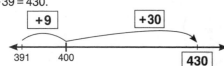

44. $193+22=\textbf{215}$.

45. $496+56=\textbf{552}$.

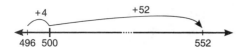

46. $598+223=\textbf{821}$.

ADDITION
Easier Sums
72-73

47. To make adding $39+26$ easier, we take 1 from 26 and give it to 39.
So, $39+26=\boxed{40}+\boxed{25}=\boxed{65}$.

48. To make adding $290+86$ easier, we take 10 from 86 and give it to 290.

$290+86=\boxed{300}+\boxed{76}=\boxed{376}$.

49. $52+128=\boxed{50}+\boxed{130}=\boxed{180}$.

50. $78+96=\boxed{74}+\boxed{100}=\boxed{174}$.

51. $393+128=\boxed{400}+\boxed{121}=\boxed{521}$.

52. $595+237=\boxed{600}+\boxed{232}=\boxed{832}$.

53. After Marie pours 4 ounces from one pitcher into the other, one pitcher holds $96+4=100$ ounces of water, and the other holds $96-4=92$ ounces of water.

The total number of ounces in the two pitchers is $100+92=\textbf{192}$ ounces.

This is the same amount of water that Marie started with, so $96+96=192$.

54. After 3 cars drive from the upper lot to the lower lot, there are $203-3=200$ cars in the upper lot, and $277+3=280$ cars in the lower lot.

So, there are $200+280=\textbf{480}$ cars in the whole garage.

This is the same number of cars that were in the garage before the 3 cars moved. So, $203+277=480$.

55. The top shelf currently holds 93 books. Adding 100 to a number is easier than adding 93. If we move 7 books from the bottom shelf to the top shelf, there will be $93+7=100$ books on the top shelf. So, we move **7** books to make it easier to find the total number of books.

Adding 100 books to the $78-7=71$ books left on the bottom shelf gives $100+71=171$ books.

56. To add $98+101+101$ peas, we take one pea each from Ada and Parker and give both to Sander. This leaves 100 peas on each plate, for a total of $100+100+100=\textbf{300}$ peas.

— *or* —

Adding $98+101+101$ by place value gives $200+90+10=\textbf{300}$ peas.

ADDITION
Adding and Taking Away
74-75

57. Adding 38 to a number is the same as adding 40, then taking away **2**.

58. Adding 19 to a number is the same as adding **20**, then taking away 1.

59. Adding **93** to a number is the same as adding 100, then taking away 7.

60. Adding 28 is the same as adding 30, then taking away 2.
So, $133+28=\boxed{133}+\boxed{30}-\boxed{2}=\boxed{161}$.

61. Adding 91 is the same as adding 100, then taking away 9.
So, $216+91=\boxed{216}+\boxed{100}-\boxed{9}=\boxed{307}$.

62. Adding 95 is the same as adding 100, then taking away 5.
So, $256+95=\boxed{256}+\boxed{100}-\boxed{5}=\boxed{351}$.

63. 27 is 3 less than 30. So, the sum of three 27's is $3+3+3=9$ less than the sum of three 30's.

We have
$$27+27+27=\boxed{30}+\boxed{30}+\boxed{30}-\boxed{9}=\boxed{81}.$$

64. 99 is 1 less than 100. So, the sum of five 99's is 5 less than the sum of five 100's.

We have
$$99+99+99+99+99=\boxed{500}-\boxed{5}=\boxed{495}.$$

65. 298 is 2 less than 300, and 299 is 1 less than 300. So, the sum $298+299+300$ is 3 less than the sum of $300+300+300=900$.

We have
$$298+299+300=\boxed{900}-\boxed{3}=\boxed{897}.$$

66. 24 is 1 less than 25.
23 is 2 less than 25.
22 is 3 less than 25.
So, $25+24+23+22$ is $1+2+3 = 6$ less than
$25+25+25+25 = 100$.

So, we have

$25+24+23+22 = \boxed{100} - \boxed{6} = \boxed{94}$.

67. 9 is 1 less than 10.
19 is 1 less than 20.
29 is 1 less than 30.
39 is 1 less than 40.
So, $9+19+29+39$ is $1+1+1+1 = 4$ less than
$10+20+30+40 = 100$.

So, we have

$9+19+29+39 = \boxed{100} - \boxed{4} = \boxed{96}$.

68. 190 is 10 less than 200.
So, $190+190+190+190$ is $10+10+10+10 = 40$ less than
$200+200+200+200 = 800$.

So, we have

$190+190+190+190 = \boxed{800} - \boxed{40} = \boxed{760}$.

ADDITION
Review   76

69. We add 100, then take away 4:
$328+96 = 328+100-4 = 428-4 = \mathbf{424}$.

— *or* —

We take 4 from 328 and give it to 96:
$328+96 = 324+100 = \mathbf{424}$.

70. We add 100, then take away 4:
$589+96 = 589+100-4 = 689-4 = \mathbf{685}$.

— *or* —

We take 4 from 589 and give it to 96:
$589+96 = 585+100 = \mathbf{685}$.

71. Since $2+96 = 98$, we have
$702+96 = 700+98 = \mathbf{798}$.

— *or* —

We add by place value:
$702+96 = 700+90+8 = \mathbf{798}$.

72. We add by place value:
$538+140 = 600+70+8 = \mathbf{678}$.

— *or* —

To add 140, we add 100, then add 40 more:
$538+140 = 538+100+40 = 638+40 = \mathbf{678}$.

73. We add by place value:
$190+140 = 200+130 = \mathbf{330}$.

— *or* —

To add 140, we add 10, then add 130 more:
$190+140 = 190+10+130 = 200+130 = \mathbf{330}$.

— *or* —

We take 10 from 140 and give it to 190:
$190+140 = 200+130 = \mathbf{330}$.

74. We add by place value:
$295+140 = 300+130+5 = \mathbf{435}$.

— *or* —

We take 5 from 140 and give it to 295:
$295+140 = 300+135 = \mathbf{435}$.

— *or* —

To add 140, we add 5, then add 135 more:
$295+140 = 295+5+135 = 300+135 = \mathbf{435}$.

75. We add by place value:
$713+65 = 700+70+8 = \mathbf{778}$.

— *or* —

We take 5 from 713 and give it to 65:
$713+65 = 708+70 = \mathbf{778}$.

— *or* —

To add 65, we add 60, then add 5 more:
$713+65 = 713+60+5 = 773+5 = \mathbf{778}$.

76. We add by place value:
$135+65 = 100+90+10 = \mathbf{200}$.

— *or* —

We take 5 from 135 and give it to 65:
$135+65 = 130+70 = \mathbf{200}$.

77. We add by place value:
$470+65 = 400+130+5 = \mathbf{535}$.

— *or* —

To add 65, we add 30, then add 35 more:
$470+65 = 470+30+35 = 500+35 = \mathbf{535}$.

— *or* —

We take 30 from 65 and give it to 470:
$470+65 = 500+35 = \mathbf{535}$.

78. We add by place value:
$339+111 = 400+40+10 = \mathbf{450}$.

— *or* —

We take 1 from 111 and give it to 339:
$339+111 = 340+110 = \mathbf{450}$.

79. We take 1 from 111 and give it to 299:
$299+111 = 300+110 = \mathbf{410}$.

— *or* —

To add 111, we add 1, then add 110 more:
$299+111 = 299+1+110 = 300+110 = \mathbf{410}$.

80. We add by place value:
$77+111 = 100+80+8 = \mathbf{188}$.

— *or* —

To add 111, we add 100, then add 11 more:
$77+111 = 77+100+11 = 177+11 = \mathbf{188}$.

81. We take 25 from 426 and give it to 75:
$75+426 = 100+401 = \mathbf{501}$.

— *or* —

We add by place value:
$75+426 = 400+90+11 = \mathbf{501}$.

82. We add by place value:
$57+75 = 120+12 = \mathbf{132}$.

— *or* —

We take 3 from 75 and give it to 57:
$57+75 = 60+72 = \mathbf{132}$.

83. We take 20 from 75 and give it to 680:
$680+75 = 700+55 = \mathbf{755}$.

— *or* —

We add by place value:
$680+75 = 600+150+5 = \mathbf{755}$.

Sum Pyramids 77

84.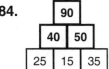

85.

86.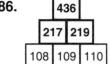

87.

88.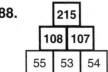

89.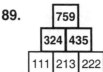

90. In the middle-right block, we have $23+\boxed{76}=99$.
In the bottom-left block, we have $\boxed{12}+11=23$.
In the bottom-right block, we have $11+\boxed{65}=\boxed{76}$.

```
        99
    23    76
  12    11    65
```

91. In the middle-right block, we have $195+\boxed{205}=400$.
In the bottom-middle block, we have $\boxed{110}+95=\boxed{205}$.
In the bottom-left block, we have $\boxed{85}+\boxed{110}=195$.

```
        400
    195    205
  85    110    95
```

92. There are no blocks that we can fill just by adding, or by finding the missing number in an addition problem.

So, we try filling in the bottom-middle block and look for a pattern. If we fill the bottom-middle block with 10, then the middle-left block is $106+10=116$, and the middle-right

block is $10+104=114$. But, $116+114=230$, not 300.

Since 230 is less than 300, we should try a larger number in the bottom-middle block. If we fill the bottom-middle block with 100, the middle-left block is $106+100=206$, and the middle-right block is $100+104=204$. But, $206+204=410$, which is too large.

We can continue increasing or decreasing our guess for the bottom-middle number until we find the number that works.

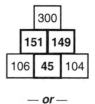

— *or* —

The bottom-middle block gets added to both 106 and 104 to give the numbers in the middle row:

$$106+\boxed{} \text{ and } \boxed{}+104.$$

We add these two numbers to get 300.

$$106+\boxed{}+\boxed{}+104=300.$$

Since $106+104=210$, the bottom-middle block is the number we add twice to 210 to get 300.

$$210+\boxed{}+\boxed{}=300.$$

Since $210+\boxed{45}+\boxed{45}=300$, the bottom-middle block is 45. We fill the rest of the pyramid as shown.

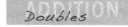
Doubles 78-79

93. We add by place value:
$34+34 = 60+8 = \mathbf{68}$.

94. We add by place value:
$38+38 = 60+16 = \mathbf{76}$.

— *or* —

We add, then take away:
$38+38 = 40+40-4 = 80-4 = \mathbf{76}$.

95. We add by place value:
$70+70 = \mathbf{140}$.

96. We add by place value:
$234+234 = 400+60+8 = \mathbf{468}$.

97. We add by place value:
$382+382=600+160+4=\mathbf{764}$.

98. We add by place value:
$390+390=600+180=\mathbf{780}$.

— *or* —

We add, then take away:
$390+390=400+400-20=\mathbf{780}$.

99. Adding four 13's is the same as adding two pairs of 13's. Each pair of 13's adds to $13+13=26$. So, we can double **26** to get the same sum as $13+13+13+13$.

Check: $13+13+13+13=52$, and $26+26=52$. ✓

100. To add $248+252$, we can take 2 from 252 and give it to 248. This gives $250+250$. So, we can double **250** to get the same sum as $248+252$.

101.

3	6	12	24	48	96	192	384	768

102.

2	4	8	16	32	64	128	256	512

103.

5	10	20	40	80	160	320	640

104.

7	14	28	56	112	224	448	896

105.

11	22	44	88	176	352	704

106. We work backwards.

First, we find the number that Ralph doubled to get 92.

$40+40=80$, so $45+45=90$, and $46+46=92$. So, the number Ralph doubled to get 92 is 46.

Then, we find the number that Ralph doubled to get 46. Since $23+23=46$, the number Ralph doubled is 23.

So, Ralph started with **23**.

107. Doubling 88 gives Winnie $88+88=176$. So, Grogg adds two 2-digit numbers to get 176. To make one of the two numbers as small as possible, we make the other number as large as possible.

The largest 2-digit number is 99. The smallest number that could be part of Grogg's sum is the number that can be added to 99 to get 176.

$99+77=176$, so the smallest number that could be part of Grogg's sum is **77**.

— *or* —

Starting with $88+88$, we can increase one of the numbers in the sum and decrease the other by the same amount without changing the sum of the numbers. For example, $88+88=89+87$.

99 is the largest 2-digit number. So, we add 11 to one number in $88+88$, and subtract 11 from the other. $88+88=99+77$.

So, the smallest number that could be part of Grogg's sum is **77**.

108. The ones digit of $12+22+32+42+52+62+72+82+92$ is the same as the ones digit of $2+2+2+2+2+2+2+2+2=1\underline{8}$.

So, the ones digit of Ralph's sum is **8**.

109. For the sum of two numbers to have ones digit 4, the sum of their ones digits must be 4 or 14. There is not a pair of numbers in the list whose ones digits sum to 4.

The only pair whose ones digits sum to 14 is $24\underline{6}+24\underline{8}$, since $6+8=14$. We circle 246 and 248.

243   244   245   ⟨246⟩   247   ⟨248⟩

110. Since we only care about the ones digit, we can ignore the 9 tens in 97 and look for the number of 7's we add to get a sum with ones digit 8.

The sum of two 7's is $7+7=14$.

Adding a third 7 gives a sum of $14+7=21$.

Adding a fourth 7 gives a sum of $21+7=28$.

$97+97+97+97$ has the same ones digit as $7+7+7+7=2\underline{8}$, which is 8.

So, **4** is the smallest number of 97's we can add to get a result with ones digit 8.

111. To find the ones digit of the sum of 111 copies of 111, we can ignore the hundreds digits and the tens digits and add all of the ones digits.

The sum of one hundred eleven 1's is $11\underline{1}$.

So, the ones digit of the sum of 111 copies of 111 is **1**.

112. 2 bananas cost $19+19=20+20-2=\mathbf{38}$ cents.

5 strawberries cost $9+9+9+9+9=50-5=\mathbf{45}$ cents.

3 mangos cost $99+99+99=300-3=\mathbf{297}$ cents.

4 dragonfruits cost $49+49+49+49=50+50+50+50-4$ cents. $50+50+50+50=200$, and $200-4=\mathbf{196}$ cents.

113. The cost of each fruit has ones digit 9.

So, the cost of 6 fruits has the same ones digit as $9+9+9+9+9+9$. To add six 9's, we can add six 10's, then take away 6. Six 10's is 60, and $60-6=54$. So, the cost of 6 fruits has ones digit 4.

The only price with ones digit 4 is **234 cents**.

233 cents   ⟨234 cents⟩   235 cents   236 cents

For example, 6 apples cost 234 cents.

114. The cost of each fruit has ones digit 9.

Since $9+9=1\underline{8}$, adding 2 fruit prices always gives a result that ends in 8.

Since $9+9+9=2\underline{7}$, adding 3 fruit prices always gives a result that ends in 7.

Since $9+9+9+9=3\underline{6}$, adding 4 fruit prices always gives a result that ends in 6.

The pattern continues. Each time we add a fruit, the ones digit of the total cost decreases by 1.

Since Winnie paid 18\underline{6} cents, she bought **4** fruits.

Note that 14 fruits also give a total sum with ones digit 6, but since Winnie did not buy 2 fruits that are the same, she could not have bought more than 6 fruits.

Winnie either bought a banana, an orange, an apple, and a mango, or she bought a strawberry, an orange, a dragonfruit, and a mango.

115. As described in the previous problem,
the cost of 1 fruit has ones digit 9,
the cost of 2 fruits has ones digit 8,
the cost of 3 fruits has ones digit 7,
the cost of 4 fruits has ones digit 6,
the cost of 5 fruits has ones digit 5, and so on.

The ones digit counts down from 9 to 0, then the pattern repeats every 10 fruits. So, the costs of 1, or 11, or 21, or 31 fruits all have ones digit 9. The costs of 2, or 12, or 22, or 32 fruits all have ones digit 8.

Since Alex paid 10<u>8</u> cents, which has ones digit 8, the number of fruits he bought is 2, or 12, or 22, or some other number ending in 2.

First, we try 2 fruits.

2 dragonfruits cost 49+49 = 98 cents, and 2 mangos cost 99+99 = 198 cents. So, 108 cents cannot be the cost of 2 of any fruit.

Next, we try 12 fruits.

12 strawberries cost 9 cents each. To add twelve 9's we can add 12 tens and take away 12 ones. 12 tens is 120, and 120−12 = 108. So, Alex bought 12 **strawberries**.

Since strawberries are the cheapest fruit, Alex could not have bought more than 12 fruits for 108 cents.

116. The ones digit of the price helps us find out how many fruits can be bought for 127 cents.

The cost of 1 fruit has ones digit 9,
the cost of 2 fruits has ones digit 8,
the cost of 3 fruits has ones digit 7,
the cost of 4 fruits has ones digit 6,
the cost of 5 fruits has ones digit 5, and so on.

The pattern repeats every 10 fruits.

Since 12<u>7</u> cents has ones digit 7, the number of fruits is 3, or 13, or 23, or some other number ending in 3.

To find the largest number of fruits that can be bought, we look at the cheapest fruit, strawberries.

23 strawberries cost more than 127 cents, so we cannot buy 23 fruits for 127 cents.

13 strawberries cost 117 cents, which is 10 cents less than 127 cents.

Since a banana costs 10 cents more than a strawberry, if we replace a strawberry with a banana, we will still have 13 fruits, but the cost will be 117+10 = 127 cents.

So, the largest number of fruits we can buy for exactly 127 cents is **13** (12 strawberries and 1 banana).

117. 88 has 5 more tens and 4 more ones than 34.
So, 34+ 54 = 88.

118. 58 has 4 more tens and 2 more ones than 16.
So, 16+ 42 = 58.

119. 91 has 6 more tens and 1 more one than 30.
So, 30+ 61 = 91.

120. 70 has 5 more tens than 24. But, adding 5 tens to 24 gives 74, which is too big. Adding 4 tens to 24 gives 64. Then, adding 6 ones gives 70.
So, 24+ 46 = 70.

121. 93 has 4 more tens than 56. But, adding 4 tens to 56 gives 96, which is too big. Adding 3 tens to 56 gives 86. Then, adding 7 ones gives 93.
So, 56+ 37 = 93.

122. 182 has 3 more tens than 154. But, adding 3 tens to 154 gives 184, which is too big. Adding 2 tens to 154 gives 174. Then, adding 8 ones gives 182.
So, 154+ 28 = 182.

123. 478 has 1 more hundred, 6 more tens, and 4 more ones than 314. So, 314+ 164 = 478.

124. We can think of 240 as 24 tens, and 160 as 16 tens. So, 240 has 8 more tens than 160. Since 8 tens is 80, 160+ 80 = 240.

125. 518 has 2 more hundreds than 333. But, adding 2 hundreds to 333 gives 533, which is too big.

Adding 1 hundred to 333 gives 433.

Then, adding 8 tens to 433 gives 513.

Finally, adding 5 ones gives 518.

All together, we added 1 hundred, 8 tens, and 5 ones.

So, 333+ 185 = 518.

126. To the right of 23 and 38 is 23+38 = 61.

To the right of 38 and 61 is 38+61 = 99.

To the right of 61 and 99 is 61+99 = 160.

To the right of 99 and 160 is 99+160 = 259.

Since 8+15 = 23, the number left of 15 is 8.

127. To the right of 81 and 131 is $81+131=212$.

To the right of 131 and 212 is $131+212=343$.

To the right of 343 and 555 is $343+555=898$.

Since $50+81=131$, the number left of 81 is 50.

Since $31+50=81$, the number left of 50 is 31.

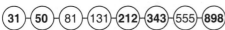

128. Since $40+65=105$, the number left of 105 is 65.

To the right of 65 and 105 is $65+105=170$.

To the right of 170 and 275 is $170+275=445$.

Since $25+40=65$, the number left of 40 is 25.

Since $15+25=40$, the number left of 25 is 15.

129. Since $201+323=524$, the number left of 524 is 323.

To the right of 323 and 524 is $323+524=847$.

Since $122+201=323$, the number left of 201 is 122.

Since $79+122=201$, the number left of 122 is 79.

Since $43+79=122$, the number left of 79 is 43.

Since $36+43=79$, the number left of 43 is 36.

ADDITION

Making 10's and 100's   84-85

130. We circle the eight pairs as shown below.

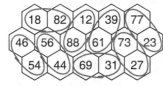

131. We only need to look at the ones digits to find pairs whose sum ends in zero.

For example, since $4+6=10$, we know $24+46$ ends in 0. We circle the eight pairs as shown below.

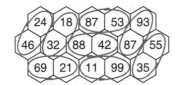

132. Each number ends in either 4 or 6. To get pairs whose sum ends in 0, we must pair each number ending in 4 with a number ending in 6.

First, we look for numbers that can be paired with only one other number. For example, 46 cannot be paired with 86 or 26, so it must be paired with 54.

84 cannot be paired with 14, 24, or 64, so it must be paired with 16.

We pair the remaining numbers as shown below.

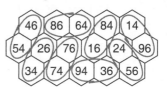

133. $40+\mathbf{39}+21=100$.
$21+\mathbf{51}+28=100$.
$40+\mathbf{32}+28=100$.

134. $\mathbf{65}+22+13=100$.
$13+\mathbf{73}+14=100$.
$\mathbf{65}+\mathbf{21}+14=100$.

135. The right side of the triangle includes two given numbers, so we start there.

$\mathbf{62}+5+33=100$.
$\mathbf{10}+28+\mathbf{62}=100$.
$\mathbf{10}+\mathbf{57}+33=100$.

136. The bottom of the triangle includes two given numbers, so we start there.

$49+32+19=100$.
$49+41+\mathbf{10}=100$.
$\mathbf{10}+\mathbf{71}+19=100$.

137. The right side of the triangle includes two given numbers, so we start there.

$\mathbf{9}+70+21=100$.
$\mathbf{3}+88+\mathbf{9}=100$.
$\mathbf{3}+\mathbf{76}+21=100$.

138. None of the sides include two given numbers, so it's hard to know where to begin.

We try placing a few different numbers at the top of the triangle, then use these numbers to complete the left and right sides so that they have a sum of 100.

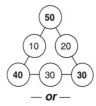

$80+30+70 = 180 \quad 70+30+60 = 160 \quad 60+30+50 = 140$

None of these work, because the sum of the bottom row is 180, 160, or 140, not 100. But, we are getting closer! Increasing the value in the top circle decreases the values in the bottom two circles. We increase the value of the top circle until the sum of the bottom row is 100.

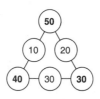

— *or* —

Whatever number we put in the top circle, the number in the bottom-left circle is always 10 more than the number in the bottom-right circle.

The sum of the three numbers in the bottom row is 100. So, the two missing numbers in the bottom row must have a sum of 70.

30 and 40 are the only two numbers that sum to 70 where one number is 10 more than the other. We use these numbers in the bottom row, and fill the top circle with 50 as shown.

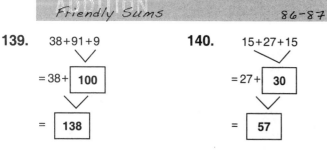

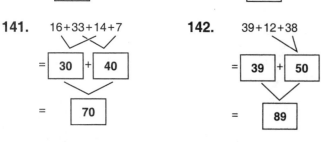

139. $38+91+9$
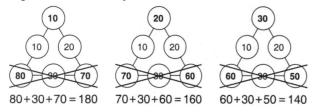

140. $15+27+15$

141. $16+33+14+7$

142. $39+12+38$

143. $49+36+51$
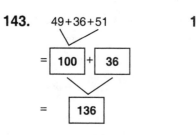
$=$ **100** $+$ **36**

$=$ **136**

144. $38+38+32+32$
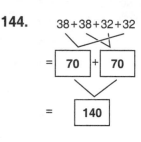
$=$ **70** $+$ **70**

$=$ **140**

145. $44+23+77+18+32$
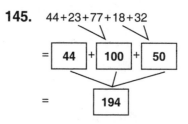
$=$ **44** $+$ **100** $+$ **50**

$=$ **194**

146. $58+20+42+39+80$
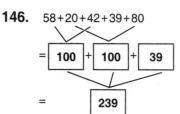
$=$ **100** $+$ **100** $+$ **39**

$=$ **239**

147. $79+21+85+15+56$
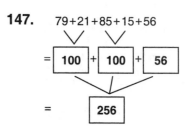
$=$ **100** $+$ **100** $+$ **56**

$=$ **256**

148. $119+226+431+74$
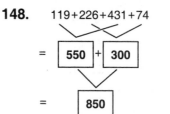
$=$ **550** $+$ **300**

$=$ **850**

149. $16+34 = 50$. We can pair each 16 with a 34 to make one 50. Then, we add five 50's to get **250**.

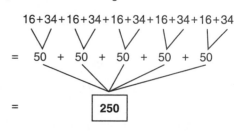

$16+34+16+34+16+34+16+34+16+34$

$=$ 50 $+$ 50 $+$ 50 $+$ 50 $+$ 50

$=$ **250**

150. We can make three pairs that sum to 40. Adding three 40's gives **120**.

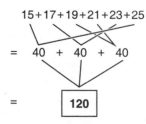

$15+17+19+21+23+25$

$=$ 40 $+$ 40 $+$ 40

$=$ **120**

151. Adding 1 through 19 is easiest if we pair numbers whose sum is 20. We can make 9 of these pairs.

$$1+19=20 \qquad 6+14=20$$
$$2+18=20 \qquad 7+13=20$$
$$3+17=20 \qquad 8+12=20$$
$$4+16=20 \qquad 9+11=20$$
$$5+15=20$$

10 is the only number we didn't pair with another number. So, the sum of every whole number from 1 to 19 is the same as the sum of nine 20's and one 10.

$$20+20+20+20+20+20+20+20+20+10$$

We add these to get **190**.

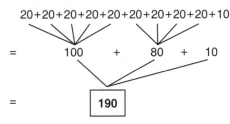

$$20+20+20+20+20+20+20+20+20+10$$

$$= \qquad 100 \quad + \quad 80 \quad + \quad 10$$

$$= \qquad \boxed{190}$$

152. Two numbers that sum to 70 have ones digits whose sum ends in 0.

2 and 8 are the only digits in the row whose sum ends in 0. So, these must be the ones digits of the numbers we circle. We find $22+48=70$.

$$1\ 1\ (2\ 2)\ 4\ (4\ 8)\ 8$$

153. If both numbers are less than 50, their sum will be less than $50+50=100$. So, one number must be at least 50.

55 and 56 are the only 2-digit numbers we can circle that are at least 50. So, we must circle either 55 or 56. We find $44+56=100$.

$$2\ 3\ 3\ (4\ 4)\ 5\ (5\ 6)$$

154. We find $654+321=975$.

$$9\ 8\ 7\ (6\ 5\ 4)(3\ 2\ 1)$$

155. We find $233+556=789$.

$$1\ (2\ 3\ 3)\ 4\ (5\ 5\ 6)\ 7\ 7\ 8\ 9\ 9$$

156. The only digits in the row are 2's and 1's.

Since 432 has ones digit 2, the numbers we circle must both have ones digit 1.

We find $211+221=432$.

$$1\ 2\ (2\ 1\ 1)(2\ 2\ 1)$$

157. The only digits in the row are 7's and 3's.

The numbers we circle must both have hundreds digit 3. Otherwise, their sum will be too large.

We find $373+337=710$.

$$7\ 3\ 3\ (3\ 7\ 3)(3\ 3\ 7)$$

158. Two numbers that sum to 796 have ones digits whose

sum ends in 6. For example, if the first number we circle has ones digit 1, the other number must have ones digit 5, since $1+5=6$.

We find $451+345=796$.

$$1\ 2\ 3\ (4\ 5\ 1)\ 2\ (3\ 4\ 5)$$

159. The only digits in the row are 1's and 4's.

The hundreds digit of the sum is 5. So, one number has hundreds digit 4, and the other has hundreds digit 1.

The tens digit of the sum is 5. So, one number has tens digit 4, and the other has tens digit 1.

The ones digit of the sum is 8. So, the ones digits of the numbers we circle must both be 4's.

We find $414+144=558$.

$$1\ 4\ (4\ 1\ 4)\ 1\ 1\ (1\ 4\ 4)\ 4$$

160. Two numbers that sum to 850 have ones digits that sum to 0 or 10. Since none of the available digits sum to 0, we look for ones digits that sum to 10.

The only digits in the row that sum to 10 are 7 and 3, or 5 and 5. We cannot circle two numbers with ones digit 5 without overlapping. So, the ones digits must be 7 and 3.

We find $117+733=850$.

$$1\ (1\ 1\ 7)\ 7\ (7\ 3\ 3)\ 3\ 5\ 5\ 5$$

161. We cannot use hundreds digit 7 or 8 in a pair of 3-digit numbers whose sum is 800. We look for a pair of 3-digit numbers with hundreds digits that are less than 7.

We find $577+223=800$.

$$5\ 5\ (5\ 7\ 7)\ 7\ 8\ 8\ 8\ 2\ (2\ 2\ 3)\ 3\ 3$$

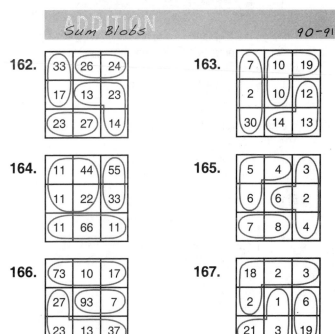

162.

33	26	24
17	13	23
23	27	14

163.

7	10	19
2	10	12
30	14	13

164.

11	44	55
11	22	33
11	66	11

165.

5	4	3
6	6	2
7	8	4

166.

73	10	17
27	93	7
23	13	37

167.

18	2	3
2	1	6
21	3	19

168.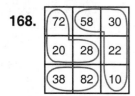

72	58	30
20	28	22
38	82	10

169.

224	324	276
276	176	424
100	524	76

170.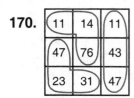

11	14	11
47	76	43
23	31	47

171.

10	20	30	29
10	10	7	14
20	13	17	20
11	12	13	14

172.

600	300	54	45
200	54	45	100
700	500	54	45
400	54	45	800

173.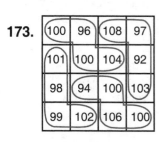

100	96	108	97
101	100	104	92
98	94	100	103
99	102	106	100

174.

22	11	22	22
33	10	11	10
22	20	11	20
44	13	13	13

ADDITION

Challenge Problems   92–93

175. Adding nine 79's is the same as adding nine 80's, then taking away nine 1's. Adding nine 80's gives a number with ones digit 0. Subtracting 9 from a number with ones digit 0 gives a result with ones digit **1**.

In fact, the sum of nine 79's is 711.

176. We can pair all of the numbers that end in 0 to get $10+20+30+40 = 100$.

Then, we add the remaining numbers to get $11+22+33 = 66$.

Finally, $100+66 = $ **166**.

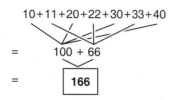

$$10+11+20+22+30+33+40$$
$$= \quad 100 + 66$$
$$= \quad \boxed{166}$$

177. We could add five copies of 222, then look for the number we double to get that sum. Since the sum of five 222's is very large, we look for a better way.

To split five 222's into two equal groups, we can put two 222's in each group. Then, we can split the remaining 222 into two equal parts. Since $111+111 = 222$, we can split 222 into two 111's.

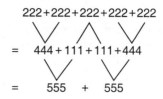

$$222+222+222+222+222$$
$$= \quad 444 + 111 + 111 + 444$$
$$= \quad 555 + 555$$

So, $222+222+111 = $ **555** is the number we double to get the sum of five 222's.

— *or* —

We can split each 222 into $111+111$. This gives us the sum of ten 111's. We can make two groups of $111+111+111+111+111 = 555$.

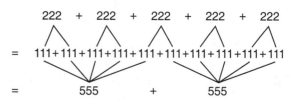

$$222 + 222 + 222 + 222 + 222$$
$$= \quad 111+111+111+111+111+111+111+111+111+111$$
$$= \quad 555 + 555$$

So, **555** is the number we double to get the sum of five 222's.

178. The smallest 3-digit number is 100. We create a list of pairs of numbers that sum to 211, starting with $100+111$.

$100+111$	$104+107$	$108+103$
$101+110$	$105+106$	$109+102$
$102+109$	$106+105$	$110+101$
$103+108$	$107+104$	$111+100$

So, there are **12** different ways to add two 3-digit numbers to get a sum of 211.

179. For a sum of three numbers to have ones digit 3, the sum of their ones digits must end in 3.

The smallest possible sum of three ones digits in the list is $5+6+7 = 18$, and the largest sum is $7+8+9 = 24$. The only number from 18 to 24 with ones digit 3 is 23, so the sum of the ones digits must be 23.

The only three ones digits in the list that have a sum of 23 are 6, 8, and 9, since $6+8+9 = 2\underline{3}$.

So, Grogg's numbers are 56, 78, and 89, and Grogg's sum is $56+78+89 = $ **223**.

180. To make the sum as small as possible, we start by making the hundreds digits of all three numbers as small as possible. We cannot write a 0 in the hundreds place. So, we use 1, 2, and 3 as the hundreds digits of our three numbers.

$$\underline{1}\,\_\_\,\_\_ + \underline{2}\,\_\_\,\_\_ + \underline{3}\,\_\_\,\_\_$$

Next, we make the tens digits as small as possible. The three smallest digits we have not used are 0, 4, and 5.

We use these as the tens digits of our three numbers.

$$\underline{1\ 0\ \_} + \underline{2\ 4\ \_} + \underline{3\ 5\ \_}$$

Finally, we make the ones digits as small as possible. The three smallest digits we have not used are 6, 7, and 8. We use these as the ones digits of our three numbers.

$$\underline{1\ 0\ 6} + \underline{2\ 4\ 7} + \underline{3\ 5\ 8}$$

Adding by place value, the hundreds digits give $100+200+300=600$, the tens digits give $0+40+50=90$, and the ones digits give $6+7+8=21$. So, the smallest possible sum of the three numbers is $600+90+21=\textbf{711}$.

Note that you could have arranged the digits differently and still gotten the same sum. As long as you used 1, 2, and 3 as the hundreds digits, 0, 4, and 5 as the tens digits, and 6, 7, and 8 as the ones digits, the sum will always be 711. For example, $148+256+307=711$.

181. To make three pairs that have the same sum, the smallest number must be paired with the biggest number. Otherwise, the sum of one of the other pairs will always be bigger. For example, if we pair 18 with 40, then the pair that includes 43 will always have a greater sum than $18+40=58$.

So, we pair 18 and 43. Their sum is $18+43=\textbf{61}$.

The other pairs are $21+40=61$ and $25+36=61$.

182. Since the sum of each group is the same, the sum of all six numbers is double the sum of each group. The sum of all six numbers is

$$10+14+21+22+30+53=150.$$

$75+75=150$, so the sum of the numbers in each group is **75**.

One group includes 22 and 53, and the other group includes 10, 14, 21, and 30.